Living the Audio Life

W0112839

Living the Audio Life details the aspects and procedures necessary for one to have a successful career in live entertainment sound. Encompassing a wide range of topics, the text clearly guides anyone interested in working in a position within the live entertainment audio field.

The guide is broken into clearly defined sections, allowing the reader to easily navigate through various subjects including jobs, career, business, creativity, lifestyle, and travel. Real-world examples and documentation from the author and key industry experts allow the reader to gain insight into the essential practices that are helpful throughout a career. Additional in-depth interviews provide details of careers from industry veterans.

Whether considering a career in live entertainment audio or just starting out, readers will find the resources for the key to success in audio. Students, those new to sound, and workers already within their careers can refer to the text as a guide throughout their journeys. With benefits to anyone interested in the audio field, *Living the Audio Life* is a key navigational resource for success.

Brad Schiller has over thirty years of experience in the entertainment field. His deep-rooted passion for the technical arts began at a very early age and his career has allowed him countless opportunities to work around the globe across diverse facets of the industry. He is also the author of *Living the Lighting Life* (Routledge, 2020) and *The Automated Lighting Programmer's Handbook*, now in its fourth edition (Routledge, 2021).

Living the Audio Life

A Guide to a Career in Live
Entertainment Sound

Brad Schiller

Routledge
Taylor & Francis Group

NEW YORK AND LONDON

Cover image: Brad Schiller

First published 2022
by Routledge
605 Third Avenue, New York, NY 10158

and by Routledge
4 Park Square, Milton Park, Abingdon, Oxon OX14 4RN

Routledge is an imprint of the Taylor & Francis Group, an informa business

© 2022 Brad Schiller

The right of Brad Schiller to be identified as author of this work has
been asserted in accordance with sections 77 and 78 of the Copyright,
Designs and Patents Act 1988.

All rights reserved. No part of this book may be reprinted or
reproduced or utilised in any form or by any electronic, mechanical,
or other means, now known or hereafter invented, including
photocopying and recording, or in any information storage or retrieval
system, without permission in writing from the publishers.

Trademark notice: Product or corporate names may be trademarks
or registered trademarks, and are used only for identification and
explanation without intent to infringe.

British Library Cataloguing-in-Publication Data
A catalogue record for this book is available from the British Library

Library of Congress Cataloging-in-Publication Data
Names: Schiller, Brad, author.
Title: Living the audio life : a guide to a career in live entertainment
    audio / Brad Schiller.
Description: Abington, Oxon ; New York : Routledge, 2022. | Includes
    bibliographical references and index.
Identifiers: LCCN 2021057455 (print) | LCCN 2021057456 (ebook) |
    ISBN 9781032037790 (hardback) | ISBN 9781032037783
    (paperback) | ISBN 9781003188957 (ebook)
Subjects: LCSH: Sound—Recording and reproducing—Vocational
    guidance. | Theaters—Electronic sound control—Vocational
    guidance. | Sonic interaction design—Vocational guidance. |
    Acoustical engineering—Vocational guidance. | Sound engineers.
Classification: LCC TK7881.4 .S43 2022 (print) | LCC TK7881.4
    (ebook) | DDC 621.389/3023—dc23/eng/20220124
LC record available at https://lccn.loc.gov/2021057455
LC ebook record available at https://lccn.loc.gov/2021057456

ISBN: 978-1-032-03779-0 (hbk)
ISBN: 978-1-032-03778-3 (pbk)
ISBN: 978-1-003-18895-7 (ebk)

DOI: 10.4324/9781003188957

Typeset in Times New Roman
by Apex CoVantage, LLC

This book is dedicated to the entire live
entertainment audio industry and all its players.

Contents

PART 6
The Travel 141

PART 7
The Interviews 159

Figures

Tables

Foreword

Before you begin reading this book, I want to address the big elephant in the middle of the room. You are probably wondering why a lighting guy has written a text about the live entertainment audio field. The simple truth is that I previously wrote a similar book titled *Living the Lighting Life: A Guide to a Career in Entertainment Lighting*, and I knew that most of the valuable information within was also relevant to an audio career. As it turns out, there are many similarities between audio and lighting professions, as we work on the same productions with many similar business situations and needs. So I felt that my knowledge and experience in live entertainment would be valuable to a wider audience than just the lighting industry.

Additionally I am fortunate that throughout my thirty-plus years in the entertainment field, I have had the opportunities to work with sound in different capacities and have also shared many Front of House (FOH) locations with my audio counterparts. In fact, my career has had many tie-ins to the audio industry in diverse ways. Starting in high school, I learned about mixing and designing sound for theatrical shows. I then worked for many years at the Irving Arts Center in Irving, Texas, as a technical director, during which I was able to work with (and specify) microphones, speakers, amplifiers, and consoles. Around the same time, I was freelancing in the Dallas area with several comedy troupes as their "tech guy." In addition to lighting and prop building, I really enjoyed mic'ing the stages, playing back music, and introducing audio in as a comedic element.

For these improvisational comedy shows I created a library of looping sound effects on leaderless cassette tapes that allowed me to respond instantly to suggestions by the audience for the various sketches. Of course, this was many years before digital playback devices were commonplace, so loading tapes quickly was the only choice. I enjoyed enhancing performances with real sounds and learning the power of the dynamics that audio can bring to a show.

When I went to work with touring and corporate productions as a lighting director, programmer, and designer, I often collaborated with the sound team to improve the production. Whether tying in timecode, receiving audio feeds, or just chatting on the bus, I found that keeping up with sound technology was very intriguing. I often made friends on the sound crew and of course enjoyed poking fun at the department, as we tend to do.

Most recently I have been employed by a lighting company called Martin that is owned by Harman, which is a sound company. Harman owns many iconic audio brands including JBL, Crown, AKG, and Soundcraft. Often I find myself interacting with my audio counterparts as we share demo spaces, trade-shows, and other customer experiences. This has allowed me to continue to have insight into the forefront of audio technology.

I have had a fantastic career in live entertainment working with a focus on lighting, but I have also known that audio was a close second in terms of a profession that always interested me. In this book, you will find a little bit of audio-specific information, but more importantly, you will find guides and suggestions for a successful career working in an exciting segment of live entertainment. I trust that, although I often refer to myself as a "lighting geek," you will see that I am dedicated to helping you have the best audio career possible and achieve all your personal goals while having fun along the way. Please enjoy the following pages of suggestions and advice catered to your audio needs and may you find success in all that you do.

Introduction

When I was about twelve years old, I was very much into magic. I actually performed at school events, parties, and more. During this time, I thought that I wanted to be an actor, but instead I discovered my passion for the technical side of the industry. Learning the ins and outs of the illusions provided me with a background for what would become a fantastic career.

My parents were very accommodating in most of my endeavors growing up, and they supported me in the majority of things in which I had an interest. After junior high school, they searched for a private school to send me to so that I could have a stronger education with better opportunities for college. Little did they know how things would actually turn out. The school we selected was based on both of its college preparatory status as well as its theater program. Remember, I was interested in being an actor. After signing up for my first acting class, I learned that I must also learn about technical theater. This is where I met Mrs. Cheryl Ellis.

This wonderful woman taught me many backstage skills, including audio and lighting. I quickly found a calling to focus on lighting and special effects. In fact, I was only an actor in one scene of my first play at the school; for the rest of that show, I was at the light board running cues! For the next three years, I was very involved with hanging and focusing lights, designing plots, running cable, and all the general crew work I could get my hands on. I would even skip my English class to spend more time in the theater. Actually, though, that did not work out so well, as I soon found that I was failing English and was told I could either leave the school or continue without theater. My parents supported my decision to leave the school.

When I entered the local public school in my junior year of high school, I was excited to move forward with my technical theater training. However, when I asked the theater teacher about the technical theater offerings, she said, "You mean the things that the people who don't get cast in the show do?" I knew right away that this program was not for me! Luckily, I discovered an amazing opportunity right across the street.

The Irving Center for Cultural Arts was an old converted movie theater that had a permanent set on the stage and a well-stocked lighting inventory. I immediately began volunteering my time to community theater groups as a lighting technician. I was thrilled to be back in the theater and growing my knowledge of lighting. Here I was working alongside both amateurs and professionals who were happy to share their knowledge with me. Occasionally the house staff would pay me to run a followspot or help with a focus.

Within a short matter of time, some of the permanent staff were moving on and I was offered a part-time position. After a few years, the other staff left, and I found myself running the entire facility! I was booking events, hiring my friends, designing shows, focusing lights, running sound, selling concessions, and everything else I could get my hands into. I was an official city employee and was making more than minimum wage while attending classes at a local community college.

At the same time, some of my friends went to work at this place in Dallas called Vari-Lite and they asked if I wanted to join too. I learned that it involved being on the road touring with bands, and I decided that I would rather stay and concentrate on theater.

I also started working with the IATSE union and a few local comedy troupes to further expand my career. I was hungry for gigs and took on as much as possible. I was fully engrossed in lighting at this point and knew it was where I wanted to be. I would read trade magazines, take classes, study the credits on television shows, and learn from those with more experience. I also attended the first-ever LDI tradeshow, which happened to be held in the same city where I lived and worked.

At that first LDI show, I came across a booth filled with Intellabeams and I learned all about High End Systems. I met Richard Belliveau and Tim Grivas and formed relationships that still continue to this day. I had a great interest in the new field of moving lights and looked for any opportunity to get my hands on them. I started programming with the earliest of controllers and taught myself all that I could. Eventually I was able to attend one of the first-ever programming seminars held by High End Systems. While continuing to work at the Irving Center of Cultural Arts, I would freelance around town as a tech or programmer.

Over time, the City of Irving built a new state-of-the-art complex called the Irving Arts Center. I took on a new position as technical director and became in charge of the smaller of the two theaters called the Dupree Theater. I was also able to attend more LDI shows and take part in industry learning on a greater scale. I never stopped freelancing as a technician, programmer, and designer. Around this time, I met Eric Durr, who owned a small number of Intellabeams and presented me with a tremendous opportunity to learn about moving light programming. My life was focused around lighting, and yet I had bigger goals to get accepted into California Institute of the Arts (Cal Arts) so I could earn a degree in lighting design.

In 1995, I packed my things and headed out to California. I was thrilled to start as a junior at Cal Arts in their lighting design program. While I enjoyed the company of other students and the opportunities available at the school, I was often asked why I was at the school. At this point I had over eight years of professional work in the industry. I had far more experience programming automated lighting than anyone at the school. While in school, I was also free-lancing around the Los Angeles area as much as possible. After one semester, I decided to leave the school to further pursue my professional lighting career.

I started working at Towards 2000 in Burbank, California under Mark Row-lands. I told him that I wanted to be a full-time programmer, but he instead offered me a job as a rental manager with the option to program when possible. I would answer the phone and send out rentals during the day, while going on gigs during nights and weekends. I never stopped learning and kept reading, studying, and meeting all the professionals that I could.

During this time, I was introduced to a brand new lighting controller called the Wholehog II. I quickly latched onto it and started to exchange emails with the creators of the desk. Within no time, I was one of the only Wholehog II programmers in Los Angeles, as well as in the country. I was not just program-ming, but also beta testing, writing libraries, requesting features, and more.

Early in 1996, Towards 2000 had the opportunity to work with Martin Light-ing to provide the first run of Martin PALs for that year's Academy Awards Ceremony. I was thrilled to be a part of this mega event as a crew chief, as I had dreams of designing the show one day. Then when the prototype Case console that Martin Lighting was using failed during pre-production, I was presented with an even better opportunity than crew chief. On the Friday before the Sun-day night live broadcast, I fired up my Wholehog II running v0.5 beta software and began programming the PAL fixtures for the Oscars. It was a thrilling experience that certainly checked a major item off from my bucket list.

Coincidently, during my time at this show, I was contacted by Tim Grivas at High End Systems. He offered me a job to move to Austin, Texas to be a part of the lighting programming team. I jumped at this chance, and a few months later, moved to Austin to start my new role.

Over the next seven years, I programmed countless productions including concerts, corporate events, theatrical shows, television productions, and more. In 2000, I was part of a team of seven programmers on the Olympic Games Opening and Closing Ceremonies in Sydney, Australia (see my diary of this time period in *The Automated Lighting Programmer's Handbook*). While working at High End Systems, I also learned about the processes of fixture and console development and assisted in the launch of many new products.

My original plan was to work for High End Systems for five years and then return to a freelance design and programming position. However, new products and major economic changes that resulted from the September 11th attacks delayed my transition. After seven-and-a-half years, I finally made the jump back to the world of freelancing. Once again, I could never have predicted how my career would transpire from this point.

About three months into my new entrepreneurial endeavor, I received a phone call from John Broderick, who was the long time LD for Metallica. We had previously worked together on many ice skating shows and he needed a programmer/operator for the upcoming Metallica world tour. After a little

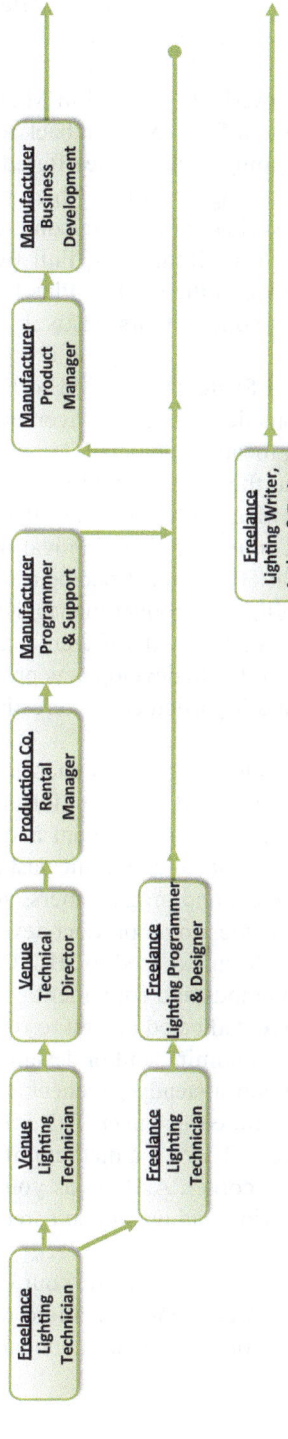

Figure 0.1 Brad Schiller's Career Path

negotiating, I found myself on a tour bus running Metallica shows alongside Butch Allen. We travelled the world for about a year and half, during which I also managed to fit in a few other freelance gigs.

Then while working the Ultra Music Festival in Miami, Richard Belliveau called and suggested that I come off the road and back to High End Systems. I had never actually considered going back, but he offered me the chance to take the lead on developing new lighting consoles (which eventually became the Hog4). So I took the job and worked with an amazing team developing many new Wholehog consoles (Hog IPC, Road Hog, Full Boar) and restoring the Wholehog brand to a prime place in the market. Plus I continued to freelance a bit on the side, programming concert tours, festivals, corporate events, and television shows.

A few years later, High End Systems was sold to Barco and my position changed to include not just console product management, but also fixtures. In my wildest dreams as a young theater technician, I could never have imagined that I could have an impact on the development of new lighting and console products in the industry. Working with a wonderful team of engineers, we developed many fixtures that continue to share innovations within the industry.

After another seven years at High End, I decided it was time to move on and took a new position at Philips as product manager for their Vari-Lite and Showline product lines. Here I worked with different teams of engineers based in Dallas and China to specify and help develop new products. As LEDs began to infiltrate the industry, I found it particularly interesting to work within this new medium.

Everything changes within big corporations such as Philips, and I found that the classic Vari-Lite that I believed in was unfortunately no longer alive. I decided to leave after two years of employment and started working with Martin as a business development manager. For the past five years, I have been the company's liaison to designers and programmers, and able to focus more on the creative aspects of the business over product development. As Martin is owned by Harman, I have also been exposed to JBL, Crown, Soundcraft and other live entertainment audio brands and equipment.

Throughout my career, I have dedicated myself to remaining involved with productions and continuing programming and/or designing whenever possible. I have been fortunate to maintain a steady paycheck for most of my career; only at a few points was I a full-time freelancer. This is one of the things I love about the live entertainment field. There is a plethora of opportunities and you never know what is around the corner. As long as you keep doing what you love, you too can have a satisfying and profitable career in the live entertainment industry.

My wish for you, the reader, is that you craft your own story with diverse opportunities throughout your career. Don't be afraid to fail or change directions, and know that prospects for success are abundant yet may not always follow your plan.

In the following pages I will present you with guidance and information to help you through your career. Much of the text is general by design as I am not in a position to provide specific legal or financial advice. However, I have consulted many in the industry to ensure that various diverse options and resources are made available to you. No matter where you are in your career, I trust that you will find some nuggets of wisdom to help you along your path.

Have fun, be safe, and may you have the same love and passion for live entertainment as I do.

Part 1

The Job

The entertainment audio field is a large industry with many different positions that one can hold throughout a career. Productions occur across a broad range of genres and allow the work to be heard in a variety of settings including concerts, theater, television, film, corporate events, nightclubs, houses of worship, and more. Understanding your role regardless of genre is a key element of a successful career.

DOI: 10.4324/9781003188957-1

The Job

1 Audio Jobs and Roles

The live entertainment audio industry consists of a plethora of vital positions that each come with its own set of unique qualities and abilities. You can make a career within just one of these roles or you can progress through many. Regardless of your path, it is important to understand the jobs that are available within the industry.

Production Audio Positions

Shop Technician

The shop technician generally performs his or her tasks within the confines of an audio or production shop. Specific duties range from warehouse attendance to show prep and equipment maintenance. In many cases, people begin their careers working in an audio or production company where they can hone their skills, gain equipment knowledge, and understand systems before participating in a crew for a show.

Show prep duties allow shop staff to work alongside production professionals as they prepare cable runs, test microphones and speakers, and sometimes even set up consoles and racks. This provides a great environment to gain valuable experience and learn from others. Furthermore, common show practices such as cabling, safety rigging, labeling, and other procedures are standard processes during shop prep.

Others will choose to specialize in equipment repair and work within an audio shop to maintain the equipment and ensure it is ready for each production. Shop technicians are an integral part of the industry as their preparation, repair, and quality control abilities have a great effect on the outcome of a production's audio.

A2/A3/Audio Technician/Stagehand

The audio technician is a very broad job category that can apply to many different aspects of production and types of personnel. In general, an audio technician is a physical labor intensive position. You might be expected to load and

DOI: 10.4324/9781003188957-2

unload trucks, set up sound rigs, run cables, wrangle microphones, organize wireless frequencies, troubleshoot situations, and much more. An audio technician is a jack-of-all-trades who should have skills that can be expanded into further career growth. Most people start their careers as an audio technician and then move on to more specialized positions.

Audio technicians are found working with all genres of sound and are essential to making a show happen. Some technicians may work as general stagehands, working with audio, lighting, video, sets, and more. Others will segment themselves to work only in audio. In either case, a good knowledge of how audio equipment operates and how a system is put together is essential. Some audio technicians are also skilled at repairing products, but this is not a requirement for most.

An audio technician must be prepared for physical labor, and in some cases, for climbing and working at heights. Understanding and following safety procedures is a must! Technicians generally are required to provide many of their own tools and follow the guidance of the crew chief, system engineer, or A1.

The audio field further classifies audio technicians into two different terms: A2 or A3 (or simply "sound number two" and "sound number three"). An A3 is a general technician that might often run mics, cables, and assist with changeovers, among others. They may also be assigned to assist with RF, systems, consoles, and more.

An A2 on the other hand is more of a primary and specialized position, and their duties are differentiated between production genres. In touring, the A2 will be responsible for many technical procedures and rig setup, while in theater the A2 is usually the person in the deck position working backstage. In television, an A2 acts as the FOH mixer's eyes, ears, and hands for almost everything happening on stage. For corporate events, the A2 will work with the microphones, RF, and is often the sound system tech or second mixer.

More Technical Positions

Audio technician roles often take on different titles or descriptions depending upon the type of production or location in the world. Table 1.1 lists many of these positions.

A1 / Mixer

Often considered the pinnacle position in the industry, the FOH mixer is responsible for operating the primary console and mixing the live sound for the audience. Equally important is the monitor mixer that will support the talent directly in hearing one another as well as in any playback. Both types of mix engineers will meet with the band or talent before a show to discuss the sound goals. Then during sound check, the mix will be adjusted according to these parameters.

Table 1.1 Audio Technician Positions

A1	An expert engineer, often a designer and/or mixer.
A2	A technical audio specialist.
A3	A general audio technician.
Crew Chief	Someone responsible for the entire sound department goals and staff.
RF Tech	A specialist of wireless system installation and operation.
Stage Tech	A technician responsible for on-stage audio needs.
PA Tech	A person responsible for installing and maintaining the PA system.
Delay Systems Tech	A specialist in setting up, operating, and maintaining an audio delay system.
Overnight Systems Tech	A general technician that is employed for overnight shifts and not the main production.
Mic Wrangler	Someone who manages microphones and maintains headset systems.
Desk/Patch Engineer	Someone responsible for configuring the console and components.
Audio Programmer	A specialist that can operate various hardware and software systems and configure them for the specific production needs.
Sound Supervisor	A specialized position-holder responsible for ensuring sound is processed correctly for televised productions.

During the performance, the mix engineers will continue to make adjustments to the audio on the fly to ensure that both the performers and the audience hear a proper mix. Mix engineers need to have a great understanding of the quality of audio, volume, balance, and EQ, and must know how to listen for sound nuances during a show.

In many production environments the FOH mixer or A1 may be responsible for the entire audio system design and construction. They must understand not only how to mix a show, but also how the audio rig is put together, how it is transported, the budget allowed, the associated costs, and much more. They may also be the lead for the audio department, meeting with artists, producers, and crew members often to discuss the sound of each performance.

System Engineer

The audio system engineer is responsible for creating a safe and effective sound system while ensuring that the mixing engineers have what they need to be successful. Part crew chief, the system engineer will design the power and rigging configurations for the sound equipment and oversee the load-in and load-out.

Once everything is set up, the system engineer will continue to tune the system to create good sound within the acoustics of the particular room. An understanding of configuration software, frequency responses, and general pressure waves is essential for any system engineer. In cases where delay speakers are utilized, the systems engineer will also calculate and configure the proper delay settings.

Sound Designer

Sound designers are most commonly found in theatrical, television, and corporate productions and are responsible for everything the audience hears. The sound designer will create sound effects, atmospheres, and sonic textures to help facilitate natural and abstract environments, which aid the show's story. They may also provide sounds related to props or other offstage effects. Usually the sound designer will select, edit, and mix audio recordings to best fit the needs of the production. In some cases they will work with a composer to create original music. They will also decide when an effect or a musical piece is played, how long it is played for, and where the sound will emanate from.

The sound designer also must understand how the audience will hear and perceive the sound and thus will make configuration changes to the acoustics of the venue or set and specify microphone types and placement locations. The ability to design the entire sound system and related equipment specifications is the essential skill of any sound designer. Most sound designers are proficient with various editing software platforms and own their own equipment as well.

Some sound designers take on multiple positions, such as mixer and/or system engineer. In some smaller markets, the sound designer may be a jack-of-all-trades and even be a crew chief and/or technician on the same production. The sound designer is also a liaison with the rest of the production team, artists, directors, band members, management, and more. Many designers will rely on a team of additional audio crew to help achieve the desired sound, while the designer is busy meeting with all the other stakeholders in the production.

Assistant/Associate Sound Designer

The role of an assistant or associate sound designer is to work alongside the sound designer and help as needed. Duties may include organizing cues, finding sound effects, creating and organizing paperwork, fetching coffee, attending meetings, choosing system components, and more. Depending upon the assistant/associate sound designer's skillset and relationship with the designer, the tasks can run the gamut from the mundane to the highly creative.

Assistant/associate sound designers hold a coveted position as they often get to work alongside the sound designer, picking up valuable knowledge, resources, and contacts along the way. Typically an assistant is on the lower rung of these two positions, whereas the associate is a closer collaborator to the sound designer. The associate will contribute more to the sound of the production and have greater responsibilities than the assistant. However, depending upon the genre, the terms may be used interchangeably. Typically in theater the roles and labels are well defined, but not so much in touring, corporate, and house of worship environments.

Account Sales

An often-overlooked position within the entertainment audio industry is that of sales. There are two key types of salespeople within the field. First, there are sound or production company salespeople who work directly with productions to help them achieve their vision, utilizing the equipment on hand and within the specified budget. Quite often they are referred to as "account reps." Then there are salespeople who work for manufacturers or distributors who negotiate deals with sound and production companies to ensure that they are purchasing the latest equipment on the market (see Audio Manufacturer Positions).

The rental or production company salesperson must have a good working knowledge of production, as this person will often make suggestions regarding gear, trucking, staging, crew, and more to assist the audio team before a production leaves the shop. Most account reps have a production background, having worked as technicians, crew chiefs, or mixers on many shows.

Most mixers or sound designers have their favorite salesperson at various sound shops and this relationship-based position is key to productions coming in on budget and on time. A good rental or production salesperson will also be the production's point person regarding all the sound equipment and staff. Often the salesperson will be on site during pre-production to ensure everything moves smoothly and to assist the sound team with any last-minute needs they may have.

Instructor/Educator

As with any field, there is a strong base of entertainment audio professionals who choose to teach and share their craft with others. Many educators in different programs, including high schools, colleges, and private schools, as well as independent educators serve an extremely useful function within the industry. Generally the instructors have a vast amount of real production experience, and most are still working with productions while teaching.

Audio Manufacturer Positions

Audio equipment manufacturers are large businesses employing many people across various disciplines. There are multitudes of ways a person can become a part of the entertainment audio industry while working with an audio equipment manufacturer. All the typical positions held within any company are available (e.g., HR, accounting, legal), but only a few positions are really specialized for the entertainment audio industry.

Product Development

Electrical, mechanical, and acoustic engineers, as well as project managers, computer programmers, product managers, and more are required to create and develop new audio products. Many of these positions are often filled by staff who have never been backstage at a show, or even have a full understanding of

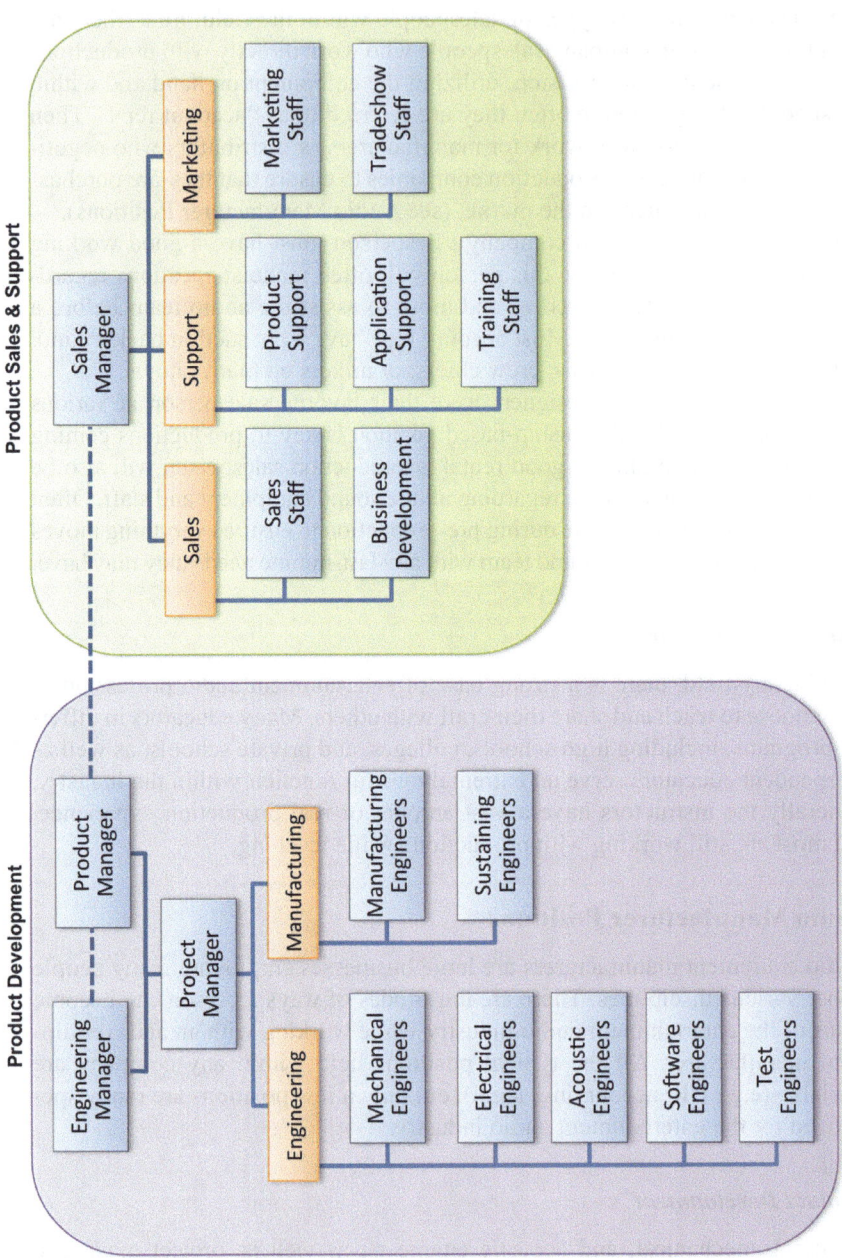

Figure 1.1 Audio Manufacturer Positions

what it takes to put up a sound system. However, their specific engineering backgrounds enable them to work within the field to create exciting new products.

Equipment Support

Audio equipment manufacturers must have a strong support team to assist productions when equipment fails, and to train the industry professionals on the operation and repair of their products. Technical support staff will be experts of their particular products, and should be able to describe solutions to nearly any problem that develops. In addition, they often teach training classes to those working in the industry. Some courses explain how to program consoles, set up speakers, and work with processing equipment, while others are deeply technical repair programs.

Many support personnel working for manufacturers have a background with numerous years of working in live audio productions in various positions. Their real-world knowledge and their product-specific familiarity make them extremely valuable resources for anyone using specific equipment. After several years, many of these product experts move on from the manufacturers to become freelance technicians, mixers, and sound designers.

Application Support

An offshoot from typical product support, an application support department will help productions directly by providing manufacturer staff to assist with various aspects of an audio production. From assisting with system layout and usage to understanding components and accessories to order, they provide an invaluable resource for customers. Often these experts will be on site at pre-production and work alongside the show's staff to properly utilize the manufacturer's equipment as required by the production. Similar to traditional support personnel, application support staff members usually have strong backgrounds working within the audio field.

Business Development

Business development roles are typically held by industry veterans who are experts across many genres of the field. They will have strong relationships with a vast number of audio professionals, as well as a robust understanding of production processes and procedures. Their core purpose is to be a liaison between the sound designers, mixers, system engineers, technicians, and the audio equipment manufacturers.

The business development staff members will also ensure that those making decisions about products in the industry are aware of the products offered by their particular manufacturer. Furthermore, they will work with the sales team, securing specifications of products and increasing product demand, which leads to increased sales.

Sales

Every equipment manufacturer and distributor has a sales team that is tasked with selling new products into the field. Because most shows do not purchase their own gear, these salespeople must interact with the sound rental and production companies. Many of the manufacturers' sales staff may not have worked on actual productions, but instead have strong backgrounds and training in sales.

However, they must quickly learn the procedures and particularities of the industry in order to properly sell their wares. Furthermore, they must be experts with their company's particular equipment and understand the needs of the sound designers, mixers, system engineers, and other audio staff.

Marketing

Every manufacturer needs to have a strong marketing department to assist in promoting their products and explaining their usage. Many will hire industry experts who have strong marketing backgrounds as well as a vast "black book" of audio contacts. These specialized marketing people will work directly with sound designers, mixers, system engineers, and others to promote their work that utilizes the manufacturer's products.

Furthermore, they will often assist in industry tradeshows and need to have an understanding as to how audio productions operate. Many will form strong relationships with mixers, sound designers, and production or sound company staff.

Many More Opportunities

There are countless more work opportunities in the field of entertainment sound. This includes common positions tailored to the industry such as marketing and press officers, accountants, warehouse employees, strategists, consultants, and more. Each serves a particular purpose with a focus on audio. Throughout your career, you can easily move between many different positions if you wish.

2 Audio Genres

The entertainment audio industry has many facets within which you can find a home in one or work through many. Most in the industry will find that they work across several genres, but specialize in one or two. For instance, a concert FOH mixer may engage in some corporate and festival work but never work in television or theater. There is great crossover of skillsets and procedures between the various genres, but the nuances of each are important to understand.

Theater

Many people start in entertainment audio through theater. It is, after all, where all this started, and the very root of the industry. Theatrical performances occur in various venues and include plays, musicals, readings, and performance art. Generally speaking, the theater is often the lowest paying genre for audio professionals. The top echelon of theater is of course Broadway and the West End, and some sound-focused practitioners can make a good living working in these environments.

Typically, theatrical production is a bit slower-paced than others, while also being highly detail oriented. Theater performances are usually temporary and the magic of the craft can be an allure for many. Theater can be held within many different types of venues and theatrical productions can also tour.

Corporate/Industrial

Many businesses present staged events such as product launches, press releases, sales meetings, shareholder events, and more. These can range from small events in a ballroom to stadium-sized productions. The genre also includes tradeshow booths, video presentations, showrooms, and other business-focused production environments.

Entertainment audio is usually a large part of these events, requiring workers in all positions imaginable. Generally corporate events have larger funds available for production and salary budgets. Many will also generously pay overtime during long production hours. In addition, the actual shows cross bounds

DOI: 10.4324/9781003188957-3

between theater, concert, television, and more. Corporate events are wonderful learning environments and can be a single genre for one's entire career.

Touring

One of the great standards in the audio industry is that of touring. Touring commonly refers to concert touring, but there are also many other touring-style events. By definition, a touring production is designed and constructed to move from one location to another, while presenting a similar performance in multiple locations. The logistics of a sound system that must be assembled, utilized, disassembled, traveled, and repeated is unique to this segment of entertainment audio. Many procedures are unique due to the very nature of touring.

Touring of musical acts is the most common form, but touring can also encompass theater, corporate, houses of worship, television, and sporting events. Whenever the same production occurs in multiple locations, utilizing much of the same gear and staff, it is considered to be touring.

House of Worship

Some form of entertainment audio augments religious ceremonies, services, events, performances, and rituals. Many of the same job positions and principles used in other genres apply with house of worship productions. However, there is an almighty purpose for the event that moves beyond a standard performance or presentation.

The audio opportunities in this genre range from spoken word to music and large-scale productions. Thus house of worship audio shares fundamentals with theater, corporate, touring, television, festivals, as well as other audio genres.

Festivals

Music festivals have become very popular over the past 50 years. From the classic Woodstock to today's EDM festivals, this type of event often includes multiple stages with various acts performing across several days. Audio mixers, system engineers, and technicians must work with different artists, managers, locations, and more while remaining focused on the overall event goals.

Festival sound rigs differ from those of touring as they are constructed for a short period of time, yet need to remain flexible for various acts. The console and inputs may need to be reconfigured multiple times during the event. Audio professionals working on festivals must understand the unique stresses applied to working with a temporary sound rig and multiple acts.

Special Events/Parties

This category is rather broad as many events can be lumped into this category. However, there are special events and parties happening all over the world at this moment, and most of them are utilizing entertainment audio. Almost every sound company, from small to big, will take part in these types of events. Whether it is a wedding, bar mitzvah, record release, fashion show, after party, or gala, these events are typically one-off occurrences.

In some cases, very little pre-production will be applied, and a truck of gear and a crew will appear to make the event happen. In others, the event will be expertly planned and prepared. The event industry can be very fast-paced, with multiple changes and decisions coming in from different directions and players. Audio professionals need to learn to handle the environment and ensure the best event possible given all the parameters and stakeholders.

Nightclubs

Nightclub sound runs the gamut from a simple DJ rig to full productions. Some venues will employ sound professionals, while others will simply have a DJ or staff person playback music. Nightclubs can present great opportunities for anyone starting out to hone their audio craft. With larger clubs and venues, a successful career can be had with opportunities to work with various acts. However, most people tend to move on from club sound to other genres.

Theme Parks/Cruise Ships

Many facets of the industry fall into a far-reaching range of entertainment with varying forms of production. Two common areas are theme parks and cruise ships. Entertainment audio professionals will find that these venues have production types that cover the extent of the industry. There are opportunities to work in any position and nearly any of the genres within a single theme park or cruise ship.

Television

Television production provides opportunities for a hybrid form of working with sound. While there is often a live audience, the needs for the broadcasted audio takes precedence as it is key for the larger audience watching the broadcast from elsewhere. Mixers and engineers need to find the balance that works best for the audience as well as the local band or performers. While many audio processes for television are identical to other forms of live entertainment, there are very unique situations and requirements that must be understood. Mixers and engineers must adhere to these differences to ensure that the audio for the broadcast meets the requirements and standards of the production.

3 Audio Skills

First and foremost, to work in the live entertainment audio field, you must have a general understanding about audio. Regardless of your experience level or current role, you need to be familiar with why and how sound is used in the live entertainment field. Even the person in the shop loading a truck with truss and road cases should have a basic knowledge of how the gear is going to be used within the production.

There are, of course, people pushing road cases and hanging PA who have no idea about sound, but generally this is not a good idea. If you plan to make a career within the industry, then it is imperative that you learn all you can along the way. If you are starting out working in a shop, ask questions, read books, watch videos, take classes, and absorb as much as possible about the business.

Early in your career you will be a generalist, studying various facets of sound. By learning as many different aspects of audio as possible, you will be able to hone in on which path you wish to take going forward. Then once you are ready to focus in on a particular position, start learning as much as possible to become a specialist in that area.

Once you narrow down your goals, keep moving forward but learn all you can about your desired position. For instance, if you wish to become a sound designer, start looking at current designs and try to recognize the designer's choices. Don't criticize, but instead question and learn. Practice at every opportunity with actual consoles, microphones, and sound editing software. Ask questions and become a sponge regarding sound design information. Most importantly of all, don't be in a hurry to hit it big. As with any career, large successes come after long periods of experience and learning. You probably won't jump overnight from hanging PA to being the sound designer or mixer of a major production.

The Best Tech

Most people in the industry start out as audio technicians in school or just out of school. During this time, they learn to hang PA, configure systems, roll cable, push road cases, and much more. The physical labor is key to gaining a working knowledge of entertainment sound. No matter where your career takes you, these experiences will assist you in your duties and responsibilities later on.

DOI: 10.4324/9781003188957-4

For example, mixers and engineers who have hung PA and loaded trucks design rigs that tend to work better on the road than those who have not. In the same manner, the best sound manufacturer engineers worked in theater or concerts while earning their engineering degrees.

Audio Skills

Sound is a physical part of the world and you should take every opportunity possible to learn and understand about audio as it pertains to your current job and your career path. You cannot expect to mix the sound in a church and then a year later be the FOH audio engineer of a major television show or concert tour. Table 3.1 details the essential elements of audio of which everyone in the industry should possess at least a basic knowledge. Depending upon your position and goals, you may choose to become a specialist in certain aspects, or even as many as possible.

Table 3.1 Audio Skills to Learn

Microphone Types and Purpose	Determining which mic is best for each voice or instrument and understanding the correct placement. Technical knowhow includes gain settings on the pre-amp and knowing when phantom power is required.
Speaker Types and Purpose	Different speaker types, locations, and setup result in different sounds. Selecting proper speakers and layout is essential.
Rigging PA and Hanging Microphones	PA rigging can be via chain motors over people's heads or speakers on tripods. Microphones are often also hung above performers or stages. All equipment placement requires perfect execution as well as attention to detail and safety.
Console Mixing Techniques	Standard procedures and terminology, programming flow, and mixing procedures. Basic console procedures and syntax along with patch information.
Networking and Protocols	Networking basic configuration and troubleshooting along with an understanding of Dante, AES67, Q-LAN, and others.
Design Aspects	Sound's emotional influence on an audience, direction of sound, and the impact on a performance.
Music Theory	The basics of tuning, timing, and harmony. Also includes melody, scales, rhythm, tonal systems, intervals, composition, and orchestration.
Timing	Understanding musical composition and theory and working with time-based systems such as SMPTE or MTC.
Editing, Drafting, and Visualization	Computer programs, standard procedures, communication, concept versus reality. Working in a 3D realm.
Frequencies and EQ Techniques	Understanding frequencies is key to optimizing the sound system and sculpting individual inputs.
Signal Flow	Knowing how the signal moves through the system and the console in order to achieve good sound and quick troubleshooting.

(Continued)

Table 3.1 (Continued)

Gain Structure and Gain Staging	As a signal flows through a system the level should be strong enough to minimize noise but not so strong that it causes clipping and/or distortion.
Dynamics Processing	The process of using hardware or software to alter the dynamic range of an audio source.
Metrics of Sound	Measuring and comprehending decibels, frequencies, compression rations, and other measurements.
Safety Procedures	Every job should look out for safety of all concerned. Recognize standard safety procedures regarding designing, hanging, connecting of equipment, as well as handling emergency situations. Be aware of dangers due to loud sounds or lack of sound.

Learning Resources

When you have the drive to learn about audio or increase your knowledge about sound, there are a vast number of resources available to you, even from young age. Some schools teach audio skills from junior high and onwards. Many colleges have programs dedicated to live entertainment sound design and/or audio mixing. Numerous production and sound companies have internship programs allowing you to simultaneously learn and earn. Plus, there are numerous books, seminars, videos, articles, and other opportunities for you to fill your audio sponge with information. Table 3.2 explains some of the most common resources for learning and enhancing your audio skills.

Table 3.2 Audio Knowledge Resources

Books	Several publishers publish works related to entertainment audio. A sample includes: Focal Press—www.routledge.com/focalpress Entertainment Technology Press—www.etbooks.co.uk
User Manuals	Every product comes with instructions. Reading user manuals will fill your brain with valuable information. They are mostly found in PDF format on manufacturer websites and as a help file in software.
Industry Press	Much of the industry press provides articles and other resources related to learning about lighting. A few include: FOH—www.fohonline.com Lighting & Sound America—www.lightingandsoundamerica.com Lighting & Sound International—www.lsionline.com Mix—www.mixonline.com Technologies for Worship—www.tfwm.com
Schools	High schools, colleges, universities, and trade schools around the world provide live audio programs.

Internships	Production companies such as OSA, Clair, and Clearwing, as well as manufacturers such as JBL, DiGiCo, and Shure offer paid internships.
Manufacturer Trainings	Most audio manufacturers offer training sessions detailing their specific products. Additional training resources are usually available on their websites too.
Tradeshows	Tradeshows are great places to learn about new products as well as to attend learning seminars and workshops. Many different entertainment-audio-focused tradeshows are held around the world. Here are a few:
	NAMM—www.namm.org
	USITT—www.usittshow.com
	ProLight and Sound—pls.messefrankfurt.com
	InfoComm—www.infocommshow.org
Online	Numerous professional and amateur websites offer information in the forms of blogs, videos, podcasts, and more. Product tutorials and demonstrations are often found on manufacturer websites.
Seminars and Workshops	Manufacturers, audio/production companies, unions, and other institutions host learning seminars or workshops, often for free.
On-the-Job Training	Every show is a learning opportunity.

4 Organizational Skills

The live entertainment audio industry has many positions and the majority of them rely on strong organizational skills. Whether a sound designer, mixer, technician, or salesperson, one must learn to organize documentation, signal flow, equipment, and personal business.

Documentation

Around the world, good documentation and data storage methods are key to nearly every business. Within the audio field, documentation is used not just for archival purposes, but also to convey information, share resources, and equalize communication processes. Sound designers and their associates produce the most documentation in the forms of speaker plots, mic allocations, signal flow diagrams, cue sheets, and more.

Console mixers tend to keep their documentation embedded within the console show files instead of creating additional paperwork or files. Even so, some will document all the console settings on a spreadsheet. Audio crew chiefs and technicians produce documents such as equipment lists, signal flow and patch sheets, repair procedures, shop orders, truck pack lists, and more. Sales personnel often document opportunities and contacts, while manufacturing staff have a plethora of documentation involved in the design and manufacture of new products.

Right From the Start

With all these various documents, audio professionals must find an organizational system that works best for themselves. Some prefer to build a binder per show, in which they place hard copies of all documents, signal flows, and other materials. Others turn to a total digital world and instead utilize folder systems within their computers to organize the various documents.

Whether digital or analog, you must start with yourself to get organized on a per show basis. First, create a folder or a binder with the name of the production. Within this parcel, make subfolders or tabs to segment the information by type. Some common examples include: plots, signal flow, patch, cues, show files,

DOI: 10.4324/9781003188957-5

sound files, shop orders, scripts/show flows, staff info, procedures, timelines, and travel documents. Fill each segment with the appropriate information and update as required. Place out-of-date versions of documents in further subfolders labeled with the word "old." Doing so will keep the most current documents in the main folder, but still allow you to access the previous versions if required.

This method will allow you to quickly and easily find any type of document per production. In the similar manner, you should organize your own business documentation. I have a folder on my computer titled "Invoices," which then has subfolders by year. Each year has copies of the invoices I have created and sent to my clients. I also have folders with yearly subfolders for my corporate business documents, such as tax forms, letters of agreement, receipts, etc.

Email communication moves the world; so it is important that you can sort through your emails to determine what is currently significant and find older information as well. A good plan to utilize folders and maintain your inbox is essential for a structured email system. I personally have email folders

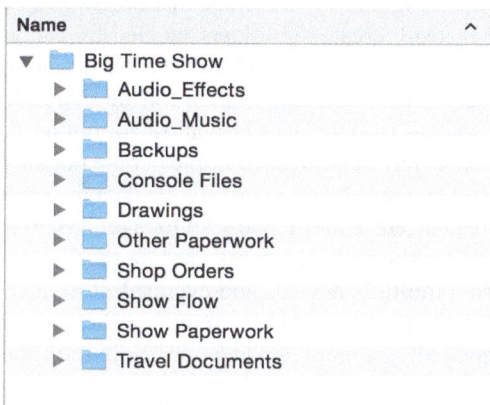

Figure 4.1 Folders for Production

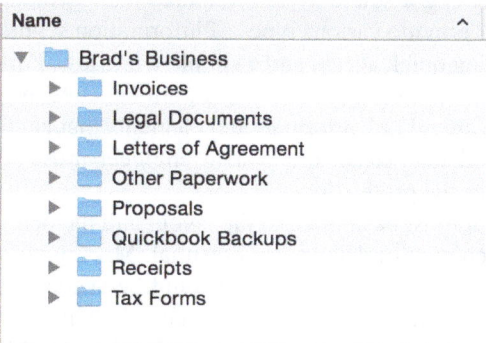

Figure 4.2 Folders for Business

designated for production mails (including subfolders by production), personal emails, travel documents, receipts, business, and more. You need to be diligent in moving emails into folders as they come in; otherwise, your inbox will become an overloaded pile of mish-mash that you will come to regret.

Shared Organization

When communicating with others regarding the audio of a production, organization is key to simplicity and reduced confusion. Many apps and tools exist to assist with organized sharing, but they all require dedication and perseverance from the user. A structured mindset is essential for success with any of these tools. File sharing services such as Dropbox and Google Drive are wonderful resources as they serve as repositories of files that can be shared and synchronized with ease.

When using a file sharing system, be sure to label all folders clearly and use a similar structure to that described above. Don't just cram all the documents for a show in a single folder and hope that others will be able to navigate through the pile. Instead, create subfolders labeled by category and disseminate the data accordingly. Some people will even place a simple text file in the main folder that details the organization plan for the shared files.

Many services such as Google Docs also allow for online shared documentation. This can be extremely helpful, as multiple people not only have access to a document, but also have the ability to update the data as needed. I find Google Sheets (online spreadsheets) extremely useful for agendas, itineraries, inventory lists, cue sheets, microphone assignments, and more. Plus you can access the documentation from multiple devices and even select to have copies available on your device when not connected to the Internet. Then any changes you make offline will automatically be merged when you rejoin the connected world.

An Engineer's World

Most system engineers are found to have advanced organizational skills, as the simple act of creating a sound system requires management. They must create drawings that provide various types of information segmented by purpose, while also allowing quick access and a simple workflow. Furthermore, data on patching, routing, EQ settings, RF configurations, cues, microphone assignments, and more are all key informational elements that cannot just be thrown haphazardly together. Instead, the system engineer must determine the best documentation for the production.

Careful thought must be applied to the layout and the operation of faders on a console and archiving of data is also essential. Mix engineers routinely store data files on and off consoles, and this should be accomplished in a manner that is well defined and organized. This process usually consists of detailed labeling of show files as well as having methodical storage locations and folder systems.

Beyond Data

Organization is not just about data and documentation. A good plan is essential for attacking tasks, and the more thought you put into a plan, the better it will be executed. When you have many duties to achieve a goal or complete a production, first make a list of the tasks entailed. Then work from the top, attacking the next right thing to get closer to your goal.

For instance, if you are a technician or crew chief, learn as much about the production as possible and then craft a plan to get from shop order to load in and beyond. The preparation can be in your mind or written down to share with others. Proper planning is a key element of organization for any series of tasks.

Personal Organization

In addition to organizing data and documentation for productions, you should try to systematize matters in your own life as much as possible. Ideally a personal backpack, road case, or workspace should maintain a clean indispensable inventory of necessities and allow you to quickly find what you need. If you have to spend five minutes digging through your bag to find a pen or open multiple drawers on your workbox to locate a roll of gaff tape, then you need to reorganize.

Think through your daily workflow and needs, and place common necessary objects in easy-to-find locations and label drawers and pockets as needed. Be sure to return items back to their original locations and replenish things that get depleted. Look beyond your audio work to determine how you can better organize objects and data within your life. Your home, smart phone, suitcase, and car are all places to arrange and simplify. When you live and work an organized life, everything becomes easier and more efficient.

5 Working Requirements

Depending on your current position or job duty, there may be certain requirements for your employment. Degrees, experience, certification, and references are commonly essential.

Experience

It should go without saying that your past experience will be a key asset for your future employment. If you have only ever worked as a sound technician at your local church, then you probably won't get a job tomorrow as a FOH mixer for a major concert tour.

When relaying information about your audio experience, be honest and concise. In many cases, you will have worked on countless different productions and do not need to list every show when applying for a new job. Instead, clearly explain your timeframe for working in your current position. For instance, simply state, "Sound technician for company X from 2017-2022." Then when asked about specific shows, explain favorite or important productions in person. No one wants to read a long list of shows that they may or may not have heard of.

You should, however, be able to explain your duties for the various productions and the abilities you have demonstrated, as well as name people you have worked alongside with. If working as a sound designer, you should put together a portfolio detailing your work and list your recently designed productions. System diagrams, drawings, cue recordings, and videos will be expected to demonstrate your creative abilities.

References

As the business is highly relationship-based, your references will often be very important to your career. When possible, list or suggest references from individuals that you worked with in the past. Remember too that in life it is never a good idea to "burn bridges." You should try to always be amenable with all those that you come into contact with. You never know who will have input in your next job or gig. When you work with a key industry person, ask them if you can list them as a reference before you put their name on your resume. Remember too

DOI: 10.4324/9781003188957-6

that the industry is relatively small, so chances are that the people reading your references may know the names you list. Never make up or exaggerate your relationships and always give thanks to those who have helped you in your career.

Certifications

Within the live entertainment audio profession, there are a few certifications that can be acquired. Most require extreme study followed by comprehensive testing. Once you earn a certification, be sure to list it on your resume.

ETCP

Entertainment electricians can gain certification through the Entertainment Technician Certification Program (ETCP). The ETCP mission statement clarifies their goals:

> We endeavor to develop a Personnel Certification Program to the highest standards that recognizes individuals who have demonstrated knowledge, skills, and abilities in specific entertainment technology disciplines. By providing a thorough, independent assessment of aptitude, ETCP strives to enhance safety, improve performance, stimulate training, reduce workplace risk, and give due recognition to the professional skills of entertainment technicians.

Some employers will require current certification with the ETCP, while others will just view it as a plus on your resume. Either way, the training is a valuable tool for the safety of all and should be considered by anyone working in the live entertainment audio field. Several approved trainers and courses have been established and testing is held at regular intervals. Full details can be found at etcp.esta.org.

Manufacturer Training

Many manufacturers of audio equipment will offer training classes that teach attendees how to operate, rig, and configure systems. As these courses teach proper usage of the equipment, they often include a certification status upon completion. Once certified, technicians can confirm to employers that they have been trained in the proper usage methodologies specific to that equipment. As every manufacturer is different, multiple certifications may be required depending upon your job.

Manufacturers of sound consoles and processors often hold training sessions to teach the programming and operation of their equipment. Rarely do they provide any type of certification other than a "completion of course" document. This is because they only teach the specifics of their product and not how to actually apply the knowledge for a production. You should never assume that because you attended console manufacturer training that you also became certified. However, these courses will provide you with in-depth knowledge of the operation of their products.

OSHA

The United States Department of Labor runs the Occupational Safety and Health Administration (OSHA), which oversees workplace safety on many levels. Just as with any industry, OSHA guidelines must be followed in the audio field as well. Furthermore, many states, unions, and employers require OSHA training and certification by their employees. Within the live audio field this is generally the OSHA 10 and OSHA 30 programs. Learn more about OSHA trainings at www.osha.gov.

These courses cover topics such as personal protective equipment, materials handling, elevated platforms, working surfaces, fall protection, electrical safety, and emergency planning. OSHA procedures for obtaining the required information and completing the test are very well defined. Many websites—as well as unions, audio shops, industry tradeshows, and other organizations—offer official courses and testing.

Recertification

Remember that most certifications are only valid for a specific period of time. You will likely need to recertify every few years to maintain your status. Always be sure to check the terms when applying for certification and then be sure to re-qualify as needed.

Union Requirements

In order to join a union, you may have to pass a test, attend trainings, pay an initiation fee, and/or demonstrate your skills. Each union and local office will have its own set of prerequisites. Some of these will be unique to the union, while others will involve nationally recognized certification programs. In many cases, the union will cover the cost of industry certification for its members.

Degrees and Experience

Many positions within the live entertainment audio industry will require the employee to have a college degree or equivalent industry experience. In many cases, the degree requirement is not as important as actual experience. Always ask the employer if your working background will fulfill the requirement in place of a degree. While a degree can be beneficial to a person's personal development, education, and resume, it often does not lead to additional employment more than having industry experience does.

International Regulations

When working outside of your home country, you will likely need to apply for and obtain a working visa. In many cases, the production will arrange this matter for you. Under every circumstance, be sure that you know the laws when working outside of your home country. The right to work, taxation, and other labor requirements could cause substantial headaches if not followed accordingly.

Once, I was almost not allowed into a country simply because I did not have the proper paperwork. I had to pay a hefty fine at customs and then have the production reimburse me afterwards. However, the trouble continued for another ten years as my file indicated the violation each time I subsequently entered that particular country. Even though the particular show that caused the initial trouble was long over, my misfortunes persisted. Always learn and follow employment regulations when working abroad.

6 Ethics of Audio

When it comes to working in the live entertainment audio field, the basic principles of conduct and professionalism apply. However, there are also several unique circumstances and instances that should always be considered.

You Are Always Part of a Team

In the industry there are absolutely no positions that operate independently. Regardless of your role, you will find that you are always working with others. It is very important that you continuously cooperate with your teammates and keep your ego in check. Always consider that the audio is just one element of the overall production.

Remember the goal of the show and the role of sound within. Perhaps the artist or producer does not want to spend additional money for another truck of gear, or maybe the director prefers not to hear heavy bass. While more gear or strong bass might be important to you, their impact on the show is more important. Decisions such as these must be tempered with an overriding understanding of audio's contribution to the final result.

Always keep in mind that you are part of an audio team. Recognize your teammates at all times and, no matter your role, never simply claim that *you* "did the sound" on a specific show. Acknowledge your position, but give credit to your crew, the sound company, the director, the artist, and all others involved. Try to use "we" statements when talking about the sound for a show.

Ask for Help

A growing career involves continuous learning. You should always ask for help when you are unsure how to achieve or accomplish something assigned to you. There is always someone who has encountered the same situation and that is willing to assist or teach you. Seek out guidance and support whenever needed.

Some will tell you to "fake it until you make it," but I find this to be lousy advice. Too often people will take on something that they are not prepared to handle. This can lead to dangerous situations, unfair costs, and incomplete

DOI: 10.4324/9781003188957-7

tasks. If you are not sure how to accomplish something, make it known that you need assistance, read a manual, or ask a peer for advice.

Safety First

The very nature of an audio rig entails positioning thousands of pounds of electrified equipment above and around a stage. Safety for all concerned is paramount at every moment. Never allow safety to slip in favor of price reductions, time savings, laziness, or other reasons. Safety is everyone's responsibility and must be adhered to above all else, even the show.

Look at your position within the show and see where you can apply safe practices. Leave the rigging tasks to the rigging crew, but do speak up if you have concerns. In addition, think about general safety for other staff members, the performers, and the audience. Situations outside your control such as weather, terrorism, and natural disasters can all be mitigated with safety planning and implementation. Keep safety at the forefront of all you do, no matter how minor.

The Event Safety Alliance (ESA) is a non-profit organization committed to promoting "life safety first" in all aspects of event production. Their programs, trainings, education, and resources all center on safety in the entertainment industry. Learn more about safety and the ESA at www.eventsafetyalliance.org.

The Show Must Go On

Since the early 1900s, the phrase "The show must go on" has been used in the entertainment field to remind all involved that the show itself is the most important factor of any presentation. This axiom should live inside you as a reminder to always do your part to ensure the success of the show. It is amazing to see a production team come together in times of crisis to carry out this mantra.

For instance, I recall a festival I was working at that had major problems with the power. Minutes before the headlining band was to take the stage, we lost all power at FOH. Everyone on the audio and lighting teams had to come together to solve the problem as quickly as possible. Then, once power was restored, all systems had to be restarted and tested before the band could take the stage. When the band finally took the stage ten minutes later, it was amazing to realize that we all were dedicated to ensuring that the show did go on.

Remember though, that "safety first" is the most important truth and thus should always trump "The show must go on" adage. Never allow a production to continue with blatant dangers to life and limb.

It's Someone's Big Moment

In the live entertainment industry we get to work on many different types of productions, and they each have their own stakeholders. The producers, performers, and even the audience each have a vested interest in the production being the best thing they have ever been a part of.

If you are working on a small corporate event with a small sound system for a local dry-cleaner company, the show may seem boring and redundant. This should not deter you from giving it your all and making it the best show possible. Remember, the people on the stage and the people in the audience *want* to be there. Use your craft to make it the magical moment that they are expecting. Honoring that the production is "someone's big moment" will provide purpose and focus to your own audio work.

Respect Intellectual Property

Designs, concepts, ideas, music, videos, software, and private information are each property of their creator unless otherwise released. Through the course of your job you will likely be exposed to many things that you should not freely share or use without proper permission. In some cases, you may even be asked to sign a legal Non-Disclosure Agreement (NDA). This document will bind your secrecy regarding a project's details. Always respect the rules set forth within the agreement.

Common intellectual property that you will encounter includes released and un-released music, custom and stock sound effects, as well as scripts or product details. It is your duty to only utilize these items as allowed through the course of the production. When you have been entrusted with content, you must handle them accordingly or face severe criminal and career penalties. Always follow copyright laws and understand music licensing before playing back any recorded music for a live audience.

Intellectual property can also pertain to certain audio products. As in most industries, counterfeit products are occasionally available. In some markets, it is common to find knock-off sound consoles that are not from the original manufacturer. They may or may not operate using the official software. Additionally, there have been copies of microphones, processors, amplifiers, and other sound products. When presented with an opportunity to use any of these illegal products, you should absolutely refuse. Only with diligence can we eliminate these counterfeits from the field.

Always be aware of intellectual property and behave in an ethical manner. If you are unclear about the rights of the content, then be sure to ask before distributing. Always give credit where credit is due and follow laws pertaining to patents and copyrights.

Dress the Part

When working in a technical position for any production, stage blacks (all black clothing) are commonly required or at least expected. During load-in and load-out, rehearsals, and other non-performance times, the dress code may be a bit relaxed. When working a front-of-house position such as a mixer or designer, and depending on the type of show, a more presentable clothing choice might be in order. When mixing the sound for an outdoor concert you can usually wear shorts and a t-shirt, but when sitting a console during a corporate event or awards broadcast you should dress much nicer.

I know a young sound technician who was hired for a high roller's New Year's Eve party in Las Vegas. He had just come off the road from concert touring and did not fully consider the nature of the event he was working. When he arrived for work on the day of the show he was wearing jean shorts and a t-shirt, the same as he had during load-in and rehearsals. Knowing that his job during the show would be to remain backstage wrangling microphones, he did not consider wearing show blacks.

As you probably guessed, he was not well received when he arrived that evening for work. He was promptly told that he needed to wear show blacks. He had to go to a nearby store and purchase black clothing at a premium price. When working any show, you should always plan to dress in show blacks, as you may have to enter the stage or the audience areas unexpectedly. Furthermore, you can never go wrong with show blacks at any production, rehearsal, or other event situation.

I also remember a time when a technician came out to the consoles during the pre-production period of a large concert tour I was working. It was a hot summer afternoon in a covered stadium and he appeared without wearing a shirt and shoes. The designer asked him why he was walking around like this, and his answer was simply that he was hot. As you can imagine, this did not go over very well, and he was promptly fired from the tour. Always dress the part for the show and your position. Plus consider your personal safety, as being barefoot in a venue is never a good idea.

Everyone Knows Each Other

The live sound industry is made up of a very small, eclectic group of people. Word gets around rather quickly and can help or hinder your career. If you work hard and are a kind person, then good things will be said of you and your working habits. Remember that burning bridges, lying, cheating, stealing, and other misgivings will continue with you throughout your career, so don't enter into these negative practices.

Also consider that falsifying or exaggerating your experience is very easy to verify. I once read a resume of a guy who claimed he was the sound mixer for the band Metallica. Most people in the industry know that Big Mick has been the band's engineer from 1984 onward. The name on this resume was certainly not Big Mick's and thus the person was immediately disregarded as a potential job candidate.

Even if you mix for a band during a performance at your venue, festival, or corporate event, this doesn't allow you to claim to be their audio mixer. That honor only goes to those who are hired directly for that specific position. Always be clear and honest regarding your career history and experience.

Word travels fast and you want to have positive comments made about you. Honesty, politeness, and hard work are all attributes that we wish to be associated with. By living with integrity, you will be well spoken for.

Sex, Drugs, and Rock and Roll

Most industries have their share of problems due to partying and related abuses; however, the entertainment industry certainly is better known for this type of debauchery in many respects. Alcohol, drugs, sex on tour buses, and favors for backstage passes are often associated with the climate we work in. In some circles, the stereotypical "roadie" image still rings true, but it is imperative for you to focus on your professionalism. What you do on your own time is your own business (as long as it complies with local laws), but do not ever think it is acceptable to work impaired or to abuse your position.

Having stated the strict message above, I can also confirm that in many cases this is not the reality. There are those who will drink or use drugs while working on productions. And while most of the time nothing bad will happen, there could be severe consequences. If you are climbing truss, working with electricity, operating motors, or doing anything else that could affect your life or the lives of others, then sobriety during work is vital. If you are sitting at a console pushing faders or are a salesperson hosting a dinner, some inebriation may be acceptable in certain circumstances. Use your best judgment here and always err on the side of caution.

Should you fall into trouble with drugs or alcohol, seek assistance and do not worry that your newly found sobriety will cause trouble within the field. Sobriety is well respected and has been achieved by many in the entertainment field. You will find compassion and assistance from many. Some tours even offer a "sober bus" where alcohol is not permitted.

When it comes to drugs and alcohol, always use your best judgment and adhere to local laws. Never trade for backstage access and continually work as a professional. Evaluate the consequences of temptations and make the best choice possible. Remember that people are constantly watching you and that nothing in the industry is a secret.

All Access Is Not All Access

Working on a production will give you access to an exciting part of society: the entertainment world. Often you will have the abilities to enter areas or see things that the general public only wishes to experience. However, just because you are issued a badge that says "all access" does not mean that you can go into the band's dressing room anytime you want. That badge simply allows you into the areas that you need to enter in order to attend to your job duties.

Furthermore, taking photos and/or videos or sharing information about a production on social media or websites is often frowned upon. It might be exciting to be backstage standing next to a famous actor or musician, but this does not give you the right to take a photo of the moment or ask for an autograph. Your job is still your job; you are not there as a fan. There is an unspoken rule that those who are working on a show must respect the celebrity and any secrecy involved with the production.

Many corporate events may be announcing a new product or strategy. You might learn the sensitive details during pre-production or rehearsal, but must comply with the company's wishes for you to remain silent until after the show airs. In many cases, stickers are placed on smart phone cameras or recording devices are not even allowed in the venue. A true audio professional will obey the privacy needs of the production without exception.

Many audio positions require direct interaction with artists and performers. From discussing the audio mix to wiring microphones and in-ear monitors, you will likely become rather close to celebrities. Always maintain a professional demeanor and concentrate on your job duties. Try to forget the allure of their fame and simply accomplish what is required.

Sometimes opportunities will arise for you to ask for an autograph or have a special personal moment with celebrities. Even in these circumstances you should remain professional and treat the famous people the same as anyone else working on the production. Often when I am asked what it is like to meet a famous person, I just explain it is no different than working for a company and having a meeting with the management team. They are just the people at the top of the organization chart. In some rare occasions, audio professionals (usually mixers or designers) will become very friendly with celebrities. I have always found these relationships to be very personal and rarely discussed outside of very tight circles of friends and family.

7   Best Show Ever

Often when asked "What is your best show ever?" people in the live entertainment industry will answer either "The last one" or "The next one." A good working mantra is to always treat the current production as the very best one with which you will ever be involved. This mindset puts forth an attitude and drive that force you to give your all to the tasks at hand.

No matter your position or the scale of the production, you should always strive to do your best to ensure the show comes across flawlessly and easily. By giving it your all, you are saying that this show is the most important thing in your career at this very moment. Sure, you might need to complete a design for your next show, or start thinking about the microphone choices for a future event, but while on this particular production, you should remain focused.

To be elite takes dedication and discipline. You cannot just decide that you are the best audio technician and then it becomes true. You have to dedicate yourself to your craft and be disciplined at working at your best in every situation. Ask yourself often if you are working at a level that would be viewed by others as average, above average, or beyond.

If your job is to mix the music, do your finest, and offer up the best mix the band has ever had. If you are an audio technician in the shop, take your time to double check everything and clearly label road cases, speaker locations, cable runs, etc. As a mic wrangler, meticulously check all mics, batteries, labeling, and placement. The drive to make the current show the greatest ever can reside in everyone involved, no matter your position. This infectious spirit is what drives incredible success and can be found from the smallest to the largest productions worldwide.

The work may be difficult or involve long hours, but with an attitude that thinks, "This is your best show ever," you know that you will get through the tough times. The more staff members that are dedicated to this mindset, the better the actual working environment. Most stress on a production comes from interpersonal relationships among those involved. If you are striving to make this show the best possible, then you can easily resolve differences with others that are motivated to achieve the same goal. In some cases, you may need to surrender your opinion for the sake of the production. This will be easiest if you are focused on making the show the most excellent one ever. When you allow your ego to overrule the show's outcome, no one wins. Remembering

DOI: 10.4324/9781003188957-8

that you are part of a team working on this production will help you with your decisions and relations with other members of the production staff.

Doing your best does not mean filling in for other people's deficiencies or always going above and beyond what is expected. Simply giving your duties the utmost of your capabilities is what is asked as a bare minimum. Often helping others or going beyond your capabilities will result in lackluster performance of your primary job. You must be careful to not slack on your main goals in the name of assisting.

For example, imagine you are assigned to mix the audio, but you go on stage to help troubleshoot a faulty microphone. During this time, no mixing is getting accomplished, due to your absence from the console. You are no longer doing your best (in your assigned position) for the production. Sure, you are helping the show by sharing your knowledge to restore the problematic microphone, but in the same course of time many settings could have been stored into the console. Now the show's mixing will suffer as a result, when a technician who is allocated to repair the microphone could have accomplished the task instead. Always use your best judgment to determine how your actions will benefit or impair the production.

Many people in the industry often find themselves working various positions on different size productions. You may have recently been the mix engineer on a major stadium concert tour and then a week later find yourself as the A1 for a wedding reception. The stadium tour played to 50,000 fans a night with a huge PA, while the wedding reception is for only 100 guests with a modest rig of speakers. You should have the same determination for both shows to ensure that they are absolutely the best they can be for the audience. Do not come to the wedding reception gig with a big head and reduce its importance. The wedding is equally (if not more) significant to the families in attendance as the concert tour is to the band on stage. Making every production your best show ever will ensure that you are giving your best to achieve your position's goals and contributing to the show's success.

The best show ever is not an excuse to skimp on safety. Your best means that everyone involved in the production as well as the audience are able to leave the event in the same way they came in. Doing your best and having the greatest show ever does not mean that you climb up a truss in a hurry without clipping in to safety gear. The best show ever should always be equal to the safest show ever.

Those who always put the production ahead of their own goals and needs will certainly have more successful careers than those who don't. Others on the production will see and feel your dedication to the accomplishment of the show and want to have you participate in future productions. This attitude shines above others who appear jaded or uninterested in the current show. It will be very apparent who is working on the show simply to get a paycheck, compared to those who want the best show ever. When you make every show the best ever you will have a wonderful, prosperous, and fun career in the live entertainment audio field.

Part 2

The Career

A career in entertainment audio becomes more than just working various jobs for many years. Most people have a passion for the business and find that the work is a labor of love and their careers are simply a part of their larger lifestyles. Many will say that sound is in their blood and that they could not live without it. They are often amazed that people will pay them for doing something so dear to their hearts.

But simply working within the entertainment audio field requires much more than job duties. A full career requires a greater understanding of business procedures, self-promotion, continual learning, and giving back.

DOI: 10.4324/9781003188957-9

Part 2

The Career

8 A Growing Career

A career is generally described as an occupational path undertaken for a major portion of a person's life with opportunities for progress. Regardless of your career aspirations within the live entertainment audio field, it is important to remember that growth requires time and experience. No one should expect to jump right into major productions without first "paying their dues" with smaller shows and lower positions.

A typical path for an audio professional usually starts with a technician role in a sound shop and then progresses into roles that work on-site with productions. In most cases, one can move from shop technician to show technician/ A2, then to A1/monitor mixer, and afterwards to FOH mixer or system engineer, and depending on the genre, to sound designer. However, this does not suggest that any one position is any less important than another. Along your way, you should always do your best in any position you find yourself, but also look for opportunities to gain experience to help you progress towards your career goals.

Continual Growth

A common method of advancing is to work with smaller productions at a higher position while also maintaining your primary (lower) position on a different show. For instance, you might be a technician on tour, building the PA and configuring the system every day. If you have dreams of mixing or designing, then you should offer to mix the sound for the opening act. In many cases, this position will be available and may even offer some additional monetary compensation. The additional pay aside, your main goal will be gaining experience as a FOH engineer. Be careful not to let this additional task distract you from your primary role as a technician, while doing your best to gain all the experience possible.

Another way to continually grow is to take all the opportunities that come your way. Many audio professionals will design or mix sound for very small productions with extremely limited budgets and even less pay. In order to survive, these professionals also spend much of their time as technicians and/or engineers on various events.

DOI: 10.4324/9781003188957-10

For example, I know a sound designer who has aspirations of designing shows on Broadway. He has a master's degree in sound design and moved to New York City right after college. He designs as many small productions as he can, but these do little to help him pay his bills. So he also works as many events in town as possible including corporate events, galas, awards shows, weddings, and more. For these events, he started off working as a technician and quickly moved up to mixing. In many cases he also designs for these productions in addition to mixing.

He averages over one hundred shows a year just working these events. While it is not contributing directly to his Broadway career goal, he is gaining valuable experience and allowing himself opportunities to design lower paying productions in New York City. His roster of contacts from the events world has also led to further projects and shows. Additionally, he finds opportunities to visit, assist, observe, and occasionally work with Broadway sound designers. He is focused on his design career, while paying his dues and his rent!

Feeding your passion might not feed your wallet, but there are plenty of other positions in the field that will support your financial needs while you work towards your larger career goals.

A Natural Progression

As you start your career, you will most likely be a generalist with many different skills. For most, this translates into working in shops or on productions as a technician or a stagehand. One day you might be loading trucks, another day building PA systems. At other times you might bolt together truss, help prep motors, or even set up a console. Your skillset will be based on a general knowledge of many different aspects of audio for productions. During this time, you will want to take as many gigs as you can to gain experience and begin to find an area that interests you for specialization. During this phase of your career, it is advisable to say "yes" to every opportunity that comes your way.

As you move along in your career, you should begin to specialize in a specific task or a position that is of interest to you. You could decide to become a system engineer, mixer, designer, A1, salesperson, or any other position you desire. While working as a generalist, you should start to focus in on your specialty and learn, train, and practice the required skills as much as possible. You can then selectively choose work that benefits your specialization.

For instance, the sound designer in New York I previously mentioned started working any gigs he could get, then after a few years focused in on mixing and design work only. Now he turns down general crew work in favor of opportunities that allow him to mix and/or design. As you progress in your career and decide on a focused path, you need to say "no" to shows that do not suit your specialization (unless, of course, you simply need the money).

When working as a specialist instead of a generalist, it is imperative that you let all your clients know your role and the skills that you can bring to their production. It is up to you to teach them how to see you. If your goal is to be a sound designer or engineer, then when you are ready, you should only take work that supports this position. You don't want to take a gig as a technician for a week only to discover you missed out on a FOH mixing opportunity at the same time. After years of grabbing every opportunity, it can be daunting to say "no" to work. It will feel odd to turn down a week's work or to pass on a tour. In time, however, you will be much happier working in your area of specialization and following your career goals.

Setting Goals

A great career strategy is to set multiple goals for yourself many years apart. You might wish to be a technician at a shop for two years, then a production technician for five years, and then a venue role, followed by one with a design focus. You are allowed to define your own timeline, but you should be realistic and flexible. There is no way to judge how much time working as a technician prepares you for your next big career leap, but in general you should never be in a rush to reach the top. If you are happy designing and mixing the sound at the church where you currently work, then you might want to continue that position longer than planned until you are ready for another challenge.

For instance, I had a plan when I first started with a lighting manufacturer to work there for five years before returning to the freelance world and focusing on programming and design. At the five-year mark I decided to hold out a bit longer due to new products that were coming to the market. I felt that it would be best for my career to stay for another year to fully assist in the

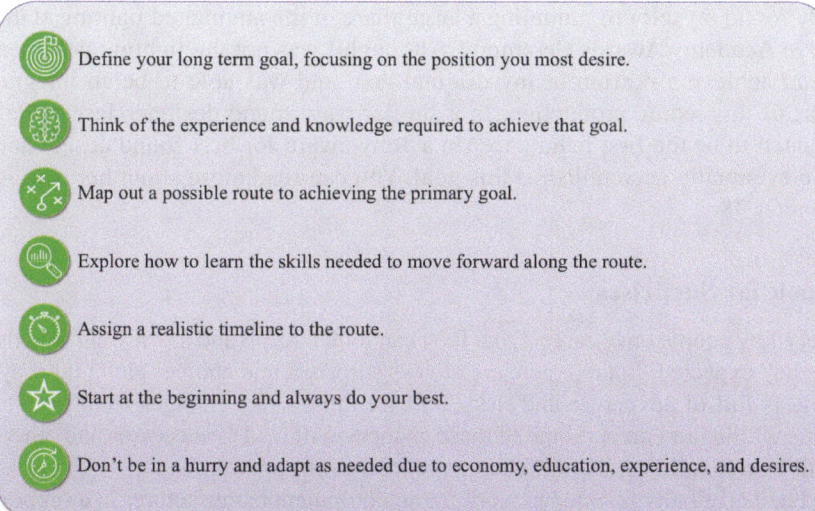

Define your long term goal, focusing on the position you most desire.

Think of the experience and knowledge required to achieve that goal.

Map out a possible route to achieving the primary goal.

Explore how to learn the skills needed to move forward along the route.

Assign a realistic timeline to the route.

Start at the beginning and always do your best.

Don't be in a hurry and adapt as needed due to economy, education, experience, and desires.

Figure 8.1 Tips for Setting Career Goals

newest technologies. Then after September 11, 2001, the economy drastically changed in the wake of the terrorist attacks. I decided to continue with the steady employment and not transition to a freelance career for one more year.

Goals are great for creating a plan, but you must be willing to extend deadlines, or make radical changes due to unforeseen forces.

A Real Example

Early in my career, I studied the lighting on the Academy Awards Ceremony (or "the Oscars") with great interest. The production was filled with fantastic automated lighting cueing and was quite a large spectacle. I recorded the show and watched segments over and over. I studied the credits to see whom the lighting professionals were that made the show happen. I had dreams of becoming the lighting designer for this production and assigned myself a goal to design the lighting for the Oscars in the next twenty years.

Now, at this time, I did not have much television lighting experience, nor did I understand all the politics and the players involved in that particular production. I simply set the goal and thought that I could work hard and achieve it. The reality I eventually learned is that the same lighting designer has lit the show for over 30 years and will probably continue for many more years. His successor will be someone with long-term television experience and strong connections in the television industry. The truth was, and continues to be, that I likely will not become the lighting designer for the Academy Awards Ceremony. My personal goal was unrealistic; I had to learn the particulars surrounding the production and the industry to realize this was the case.

Not all was lost, however! Through a series of unexpected events, I actually found myself programming a large share of the automated lighting at the 1996 Academy Awards Ceremony. Although I was not the lighting designer, I *did* achieve a portion of my original goal and was able to be an integral part of this iconic production. In a similar way, sound designer Jessica Paz wanted to be the first female to win a Tony award for best sound design and she eventually accomplished this goal. You can read more about her path in chapter 38.

Look for Surprises

Very few people have careers that they can look back at and say it went exactly as they expected. In fact, there are always surprises and choices along the way. Life is full of adventure and risks, and we must make changes from time to time within our career. Some of these changes will lead to successes and others will lead to failures; but all failures are lessons for growth.

I had originally planned to work for an equipment manufacturer in a support role for only five years before returning to the freelance world and becoming successful as an independent designer. After seven years, I did return to the freelance path and was following my plan. Then a surprise phone call presented me with an opportunity to return to the manufacturer, this time in the Engineering department helping to develop consoles and fixtures.

While I had never considered this as a part of my career plan, I did have a bit of experience in the product development field from my previous tenure. I was intrigued by the possibility of contributing to the development of world-class products used by the entire industry. I was torn, though, as my freelance career was going extremely well with lots of growth and opportunities. Ultimately, I took the job, returning with an agreement that allowed me to continue to do a bit of freelancing as well. Over the following years, I learned that I had a talent for product management and this opened new and unexpected doorways within my career.

The live entertainment audio field is full of opportunities, and you should always consider unexpected possibilities. Transitioning to companies, joining venue staff, gaining employment with manufacturers, working with only one band, taking a resident position, starting a rental company, becoming a consultant, teaching others, and developing new products are just some of the changes that sound professionals have made to their career paths. The choices you make are up to you, and no matter what you decide, you will have a great career within the live entertainment field.

Prepare for the Unexpected

Entertainment has been around for centuries and fortunately people will always want to be entertained. However, there is no guarantee that your position or work opportunities will always be available to you. You must prepare for the unexpected by saving money, learning diversified skills, and identifying when to pivot.

Shows come and go, tours stop suddenly, corporate work dries up, and work can disappear completely without notice. At any moment, your upcoming gigs could cancel. It is highly advisable to have at least six months of your living expenses in an emergency fund for slow times. You might have to tap into these resources to survive the down periods. Additionally, you may need to look for work outside the field to keep income coming in. There is no shame in taking any job if it means the difference between paying or not paying your bills.

When entertainment jobs are not available, see where you can use your skills in other industries. For instance, see if you can offer your services to machine shops, engineering offices, and other mechanical design companies. Perhaps your technical or organizational skills can be used in construction, warehousing, or even medical fields. Designers may find related work in art studios, museums, and corporate firms.

In 2020 the COVID-19 pandemic opened everyone's eyes to the fragility of the live entertainment industry. Beyond just sound, nearly every single segment of entertainment was put on pause for an unprecedented period of time.

Many lost jobs and were forced to scramble to find any possible work. Without any jobs for the foreseeable future, those who could turned to their emergency funds. However, many without any such backup plan instantly found themselves in severe trouble. With little opportunity to pivot to a new market, their suffering became unbearable. Preparing for the unexpected should always be a priority. Remember the unimaginable can occur any time without warning.

During the downtime in 2020 and 2021, many audio-related learning resources became available to the masses. Industry manufacturers, sound experts, and schools provided untold amounts of online training and workshops. This became an important time for designers, engineers, and technicians to learn new skills and improve existing knowledge. Resourceful people used the unfortunate crisis as an opportunity to come out stronger. Some even prepared for future events by learning skills outside of audio to enable them to continue working should future pauses occur to the entertainment industry.

We never know what tomorrow will bring and no show or event is ever one-hundred percent guaranteed. Take the time to prepare for the unexpected by saving money, improving skills, and understanding your value in various markets. Never take the current moment for granted and always look ahead with caution and optimism. However, rest assured that entertainment is important to the world and has always continued through the tough times.

9 Attitude Is Everything

The audio field is heavily based upon individual relationships, and thus your people skills are some of your most important assets. When I surveyed over a thousand people in the production industry for advice they wish to pass on, one theme was overwhelmingly apparent: your personal attitude and interactions with other humans are extremely important. To put it simply: if people like working with you, then you will work with more people. Never behave as a know-it-all, a smart-ass, a jerk, or an entitled person. Instead, you should be a friendly team member who does your job well. Otherwise, you likely will not get called back to work on future productions.

Sure, there are times when you need to be assertive and demonstrate your expertise about a situation or procedure. During these moments, try to present with kindness and understanding as opposed to simply forcing your opinions or knowledge on others. Safety concerns should always be expressed openly, honestly, and with emphasis. Do not hesitate to point out unsafe situations or potentially dangerous errors. Work with all the teams on a production to resolve problems instead of speaking out against others in a negative manner. Being kind should be your standard mode of operation.

Once my wife and I were at a sandwich shop in New York and ordered two vegetable sandwiches. As the guy was making them, my wife and I talked with him about New York and other topics. When he presented us with the sandwiches they were beautifully made with many vegetables and other goodies. My wife commented about how filled-up the sandwiches were, and the guy told us something amazing. He explained that not all sandwiches are made to the same standard. He said, "If you are nice you get a nice sandwich, but if you are a jerk then you get a jerk sandwich." This nugget of wisdom holds true in many aspects of life, but especially regarding your career. When you are a nice person, good things will happen in your life. If you are a jerk, then you will get what you deserve.

The vast majority of successful sound people are also the nicest. They have learned how to work well with others, keep their discomfort and stress to themselves, and navigate production politics. They know their skill level and ask for assistance when needed. They are honest with others and want the best for all involved.

DOI: 10.4324/9781003188957-11

I have also met others who have bad attitudes and seem to be stuck in their current roles or disappear completely from the industry after a while. I was once working a show in a large theater and had a clash with a stagehand nicknamed "Grumpy." He fit his moniker and acted as if he knew more than anyone and did not want to be at work. His negative attitude towards everything and everyone was a detriment to the entire production. Many on the show were having difficulties due to Grumpy's bad attitude. By the third day, the union steward transferred him to another show. Apparently, not long after that, he was told to change his attitude or find a new profession.

Some say that success in the entertainment business is based upon 40 percent skill and 60 percent personality. I tend to agree with this formula as I have witnessed it throughout my journey. Treat and respect others as you would like to be treated. Make each experience one that encourages people to want to work with you on future productions. With a great attitude and hard work, you will certainly be asked back to future shows.

Focus on Now

Your outlook will be better if you are focused on the production and the job at hand. Always stay focused on where you are at the moment and try to avoid distractions such as social media, texting, and working on other productions. When your mind wanders from the current show, your stress level will escalate and your attitude will deteriorate.

Remember that you are paid to be on-site at this production to help make it the best show ever. In return, you should give it your full attention. Do any less and you are putting your personal needs and choices above the production. Distractions may alter your mindset and place you in a negative mood. This will then carry over to your feelings about the show and the people you are working alongside. Instead, save outside influences for non-working hours and keep a smile on your face.

Sure, the rehearsals may be long, the systems engineer might ask for more changes, or the band may change their mind again. These are all stressors that can occur at any moment of a show. As they come to you, try not to say "no," and instead look for opportunities in the situation at hand. If you can take changes in stride with a focus on the end result, then your relaxed attitude will flow through to the rest of the team. Plus your "can-do" attitude will be admired and respected by those making the requests.

In some cases, time will be of the essence and work will have to happen right away. For instance, if the PA fails ten minutes before doors open, the crew will have to come together to solve the problem and get the system working again. Even if you had other plans in the moment, this emergency will require your full attention and a good attitude towards solving the problem. Although it might have conflicted with your dinner break or personal time, remembering that the show is the reason you are there will keep your attitude in check.

Thank You

Kindness and gratitude are the main elements of a good attitude. If you can be kind to others and grateful for the opportunities presented to you, then you will lead a successful career in the live entertainment audio field. Take the time to thank others daily and always follow up after a show.

Always make it a priority at the end of a show to thank each of the sound crew as well as other production staff for all their assistance. Furthermore, follow up with the people that hired you via a phone call or email, letting them know how much you enjoyed working with them. These simple touches go a long way, ensuring that people will want to work with you again (assuming you did a great job of course).

Remember that you are very fortunate to get paid to work with sound in the entertainment business. Even with the various stressors, try to bring yourself back to a state of gratitude for the opportunities to work in this amazing field. Many people spend much of their lives working purely for paychecks, but most of us are very fortunate to get paid for jobs that we love to do.

10 Training and Learning

Throughout your career you should be open to learning new information regarding sound. From technology to techniques, there is always something to learn. As your path changes you will need to study your new direction while also improving your personal skills. The industry provides many avenues of education, and each has its own benefits. Generally, a combination of all of the following is best for anyone working within the live entertainment audio profession.

High School

Many begin their entertainment audio education while in high school. During this time, basic knowledge of microphones, mixing, design, and other audio essentials is usually acquired. Some schools have simple systems with older technologies, while a few schools possess high budgets and the latest equipment.

The real key in any of these situations is the instructor. Unfortunately, it is common in high school productions for the technical aspects to take a back seat to the acting. In many cases the theater teachers are trained in and focused only on acting with very little sound or other technical background. This can lead to students learning the theater technologies on their own or from older students and parents. Look for schools that have a strong technical theater program and experienced instructors.

Colleges and Universities

College and university programs are wonderful environments to learn about audio in. Many schools provide in-depth sound mixing, design, engineering, and general production programs that culminate in BA and MA degrees. As you learn about audio, you will also receive a basic liberal arts education through a general learning curriculum. Your focus on sound will provide you with the fundamentals and basic skills to move forward with an audio career. A good school will provide many diverse production opportunities throughout the school years.

DOI: 10.4324/9781003188957-12

In most cases, colleges will offer a degree program focused entirely on sound design. While in school, you will learn audio history, design concepts, drafting, practical skills, and more. Because the focus is often on design, you may graduate lacking some of the basic audio techniques or engineering proficiencies required to enter the working world. Upon graduation you might initially find more employment prospects in these positions than in sound design. Chances are low that you will become a busy sound designer right out of school.

Countless schools are very theater-centric and may not fully relate to the plethora of live audio opportunities that you will find upon graduation. While theatrical methods certainly are foundations of most entertainment audio productions, actual "theater only" careers are rare. Some schools such as City Tech in New York City and University of North Carolina School of the Arts have stepped away from the emphasis on theater by offering degrees in production or sound technology. They utilize the latest equipment and also include audio courses that cover various genres of production such as corporate events, concerts, theme parks, and events.

Remember too that, in educational environments, the students generally learn using only the in-house equipment. This means you might miss out on the latest (or even current) technology being deployed in the field. Look for a program that is not isolated in terms of its available technology. In addition, try to select a school with a focus on relationships with outside production staff, creative uses of sound equipment, hands-on opportunities, and large budgets for rentals. Some schools will even invite well-known sound designers and engineers to teach classes, evaluate portfolios, or mix shows.

College programs are wonderful and can provide a very robust set of courses that are sure to teach anyone the fundamentals of the live entertainment audio industry. A college degree also provides many more life-enriching skills that can be used beyond the sound industry. However, this type of training can take several years and usually costs large sums of money. Completing a degree program could leave you with a large debt to repay.

When looking into a college, it is imperative that you enquire about their equipment use (in-house and rental), number of shows, visiting professionals, curriculum scope, and availability of gear for the students' usage. Ask what production positions are available to you while in school, and be clear about your post-graduation goals. Take the time to find a school with a program that best fits your desire for your future.

Technical School

Beyond traditional colleges and universities, there are several industry-focused schools with degree and certificate programs that specialize in various aspects of the field. For instance, Full Sail University (www.fullsail.edu) in Florida offers a degree in Show Production that includes courses in sound design, audio measurement systems, audio and visual technologies, and more. They even make available complete educational learning packages that provide students with a laptop loaded with important software as well as other professional tools.

Similarly, the Stagecraft Institute of Las Vegas (www.stagecraftinstitute.com) employs working professionals to teach various audio courses. They award a seal of qualification upon completion of coursework and testing. Many other technical schools can be found throughout the United States and in other parts of the world as well. Some might find technical schools expensive or limited in their training resources, while others find them a quicker path to employment than traditional colleges.

Internships

A great method to learn and hone your skills is to work as an intern within a sound or production company, theater company, or manufacturer. Most of these positions are paid, so that you can earn money as you learn. Production Resource Group (PRG) has one of the most recognized internship programs in the industry. At most of their locations, they offer a well-organized curriculum that takes attendees through multiple years of employment. As the intern works through various positions, they work with a mentor that helps guide future career choices. Upon completion of the internship, a position in the shop, on the road, or in production crews is usually offered.

Many other companies such as Clair Brothers, Clearwing Productions, JBL, and major venues provided internship opportunities to many professionals when they were starting out. The relationships formed while working indirectly for a company are usually lifelong, and many interns gain employment from their sponsoring company.

On-the-Job Training

Many regular jobs offer on-the-job training, which is similar to an internship. Training ranges from company-required courses to continued education paid for by the employer. Often, senior-level employees will share knowledge and experience with junior employees. Many venues and sound or production company positions offer learning opportunities.

Manufacturer Training

Most audio equipment manufacturers provide training sessions specific to their products. Some of these courses are required for warranty-level repairs, while others simply teach how to operate specific pieces of equipment. Often manufacturer training is provided at little or no cost to end users and can be very valuable in understanding a product or a range of products.

For instance, sound console manufacturers regularly teach courses about how to operate their desks. While these courses do not teach the procedures and skills of mixing, they do teach the features, required keystrokes, setup, and operation of their consoles. These courses are relevant to anyone working in the entertainment audio field, as they provide a basis of understanding about a specific console line. A technician may be called upon to use a desk during pre-production, while a designer might want to understand the capabilities provided to their mixer.

Manufacturers also commonly provide training classes for product repair or installation. Many of these classes are extremely technical and include testing and certification. Some will even have pre-qualifications to ensure that only the qualified people attend. These trainings are usually very beneficial as they provide insight into and skills about products that often are not available from any other source.

Manufacturer training occurs at many different locations. Often the courses happen at the manufacturer's offices, but also can be found at tradeshows, open houses, union sponsored events, and more. Always check the manufacturer's website to see what courses are available and in what locations.

Learning Sessions

Beyond manufacturer-based trainings, there are many opportunities for learning within the industry. Learning sessions, forums, panel discussions, and case studies are common parts of every industry tradeshow. Since 1901, the National Association of Music Merchants (NAMM) tradeshow has provided hundreds of training sessions annually. Many of these are available at little or no charge, while others come with a hefty price tag. Other tradeshows such as USITT, Musikmesse, and ProLight & Sound have similar training opportunities.

Learning sessions are also found at production company "open house" events, manufacturer demos, houses of worship, union halls, and even in venues. Often the hosting institution will bring in working professionals to share their knowledge and skills with the attendees. These opportunities provide valuable resources and information for students and professionals alike. Many learning sessions are also hosted online or available as recordings to stream at your own pace.

User Manuals

A great place to learn more about a specific piece of equipment is the user manual. This may seem simple, but most people skip reading the boring guide that comes with a product. However, I can assure you that if you take the time to read through the manual for a new console or processor, you will find some new information that proves valuable for your next production.

Audio equipment is very complex and each piece comes with its own software, hardware, and unique abilities. The user manual will describe all of these elements in detail and may even alert you to dangers that you did not know existed. It is always a good idea to read through a manual for any new product that you encounter. This is one of the best ways to get the most from the products you are using. Why would you want to miss out on important features or abilities just because you did not take the time to read and learn about them?

All user manuals are available from manufacturers on their websites. You can download the documents and read them on your smart phone, tablet, and/ or computer. Keep the files organized on your platform. Remember to check for updates as products and their manuals get revised over time with different software features and hardware changes. Always download and read the current manual for a product you have never used. This practice will ensure that you are aware of features and procedures related to that specific equipment.

Peer Groups

The audio industry is filled with people who are all at different stages of their careers. Everyone has something to share with others. Take the time to sit down with colleagues working at the same level as you and discuss your skills and knowledge. You will likely learn something new while sharing your own experiences. This type of knowledge exchange is always best in person, but can also occur via online meetings, social media, blogs, email, or even industry events.

Don't wait for these opportunities to come to you. Instead, take charge and invite others to meet for coffee, or ask questions at a gig when applicable. For instance, if you have the desire to become an engineer, ask the engineer on your next gig if you can watch them work. Once you have completed your assigned tasks, sit quietly at FOH and observe the process. Most people are willing to share their knowledge and will be happy to have you shadow them while they work. You can also arrange meetups away from productions to further your conversations with peers.

Private Training

There are many people with vast skills and experience in the industry. You can easily reach out to any and ask for training and education. Some may train you for no cost, while others will ask for payment. Either way, you will likely gain valuable information that may not have been available to you any other way.

Some organizations such as unions, rental companies, houses of worship, or schools may hire professionals to teach their staff important abilities. Many freelancers offer their training services and a few have even created companies focused on training. For instance, Richard Cadena specializes in various types of training with his Academy of Production Technology (www.aptxl.com). He has traveled the world teaching diverse courses to anyone that chooses to hire him.

Media

Educational materials can be found throughout the industry media as well. Publications such as Lighting & Sound America, FOH, and Mix create content that ranges from magazines to webinars. They employ industry experts to share details about productions, as well as publish how-to articles and other learning opportunities. These resources are valuable as they are usually timely and in line with current industry trends and products. In all parts of the world there are similar publications such as Lighting & Sound International, Sound & Light, Show Technology Asia, and Live Sound International.

Beyond the industry press, many online blogs, websites, YouTube videos, podcasts, and social media groups are filled with educational material. Remember, though, that not all of it is curated, and thus it is always important to consider the source. Look for sites that belong to respected industry professionals or others that have already been vetted regarding their experience and knowledge.

Books

Since you are currently reading this book, you probably already understand the importance of the printed word. Many books on the topic of live entertainment audio are available from numerous sources. Take the time to find the publications that cover the topics you need to learn and read the latest editions. Most books are now available in traditional printed format as well as electronic versions. Amazon and other online retailers make it easy to find books about live entertainment audio.

Self-Taught

I have spoken with a few audio professionals who claim they are "self-taught." However, I highly doubt that this is exactly true. While they probably mean that they never attended a school or seminar, they likely had to consult peers or read manuals to gather further knowledge. Remember that if you really want to succeed in this industry then you are going to have to go beyond yourself and learn from all the above-mentioned resources. If you limit yourself to what you can figure out by yourself, you will certainly be missing out on many key bits of information.

Learning opportunities are available throughout the industry. In order to have a very successful career, you must continue to educate yourself. Never think that you are complete in your education; instead look for learning moments to help you advance forward (or simply keep up with technology).

11 Promoting Yourself

The live entertainment audio industry is actually rather small compared to many other industries. Because of this, word of mouth and reputation are key factors to gaining employment regardless of the position you desire. While a resume is essential for some jobs, it is merely a formal step in the hiring process.

Throughout your career, you will need to continually promote yourself, your abilities, and your desire to move forward. There are countless avenues upon which this promotion can take place, and it is very important that you make use of as many as possible. You are your own brand and only you are responsible for promoting yourself. Remember that you must first do the work, and then promote yourself. Don't publicize yourself as a sound designer or mix engineer if you have never worked as one.

A Simple Resume

When building a resume for any position, it is generally advised to keep it to a single page. This is true within the audio field as well. Many try to list all the shows that they have been a part of and this just becomes a waste of time and space. Instead of listing every production, highlight a few key shows in an introductory paragraph.

Then be clear about the types of positions you have held and the skills that you possess. This level of detail is far more useful to a potential employer than reading the title of a show or name of a band. If you have worked as an A1 for twenty different tours, simply list the years you have been touring.

Education, special acknowledgments, and unique abilities should also be noted. A few references are always helpful too. Remember to keep your resume simple and to the point. It is not this document that will earn you a job, but rather your skills, personality, and references.

Portfolio

Design positions often require a portfolio that can represent your work and creativity. A good portfolio will include recordings, videos, cue lists, system

DOI: 10.4324/9781003188957-13

drawings, and more. A portfolio might be a printed document, a large display, or a virtual online experience. In some cases, designers will send their portfolios via a USB stick or on a tablet device.

Along with a resume and/or portfolio, a designer will often include a list of many of their most recent designs along with their production details such as the director, producing company, and dates. If you have been working for many years, there is no need to list all of your past shows. Instead, feature the most recent as well as a few favorites or standouts from your career.

Website

It never hurts to have a website to promote yourself and your work. It is one simple place that people can find out what you have worked on, see and hear samples of your work, and get in contact with you. There are many companies that allow you to have a free site (with some advertising), or you can choose to pay for hosting and build your own site. Furthermore, it can be helpful to register a custom URL such as www.bradschiller.com to make it easy for people to find you and remember your name.

Your website should have a clean, modern look, detailing what you do and the types of jobs you want to gain in the future. If you are a designer, having many recordings and videos of your work is essential. If working another aspect of the industry as a technician, engineer, or crew chief, you might want to include links to YouTube videos of shows that you were involved in. In all cases, be very clear about your role on the show and credit the sound designer, mixer, system engineer, and other important people and companies.

Your website also needs to have a "call to action" focus that requests viewers to contact you, read your blog, or follow your social media. If the site is merely a portfolio of your work, then visitors will be simple viewers. You want to provide them with as many opportunities to learn about you and your work as possible. A blog or video portal that shares your experience and knowledge will entice people to frequently visit your site.

Search Engine Optimization (SEO) services are big businesses in today's Internet world. They will help ensure that your webpage is chosen among the many during an Internet search. Whether you choose to hire these services or not, you need to take advantage of free SEO tools for your website. Keywords, metadata, long semantic indexing, and localization are all primary SEO principles to incorporate into your website for maximum results.

Also, don't forget to optimize your site for mobile devices. While it may look great on a computer, it needs special formatting and coding to enhance its effectiveness on other gadgets. Studies show that more searches are conducted on smart phones than on any other device. This means that the search results are also viewed on these devices. If your website does not translate well to current technology, then it will reflect poorly on you.

Don't let your website get out of date. Take the time once every few months to update the data within, and, about once a year, consider redesigning the look. You might want to remove some older show references as well as add the most recent. When you have been a part of a podcast, referenced in a review, or mentioned in a magazine or other promotional items, be sure to add links to those regularly. Furthermore, you should confirm that all links on your site are working and check that your SEO strategy is still relevant.

Social Media

A great way to promote yourself and your work is through social media. Many different and ever-changing platforms allow you to engage with others around the globe. A good strategy for promotion through social media can certainly boost awareness of you and your work among others in the field. Moreover, it can be an opportunity for you to share original content and interact with fellow audio professionals and enthusiasts.

However, simply posting updates about yourself and your work is not enough in today's busy world. You need to spend some time forming a network of followers, and then provide them with content that entices them to continue to refer back to you and your social activity. This content might include interesting thoughts, articles, or links to videos and websites. If you provide a source of valuable information, then others will want to follow you and your postings.

Self-promotion via social media should be planned and performed carefully to achieve the best results. Learn the various platforms and use them as intended. For instance, don't post sales-specific or personal content on LinkedIn, but do share recent project photos or case studies. On every site, take the time to curate your content and make certain that your photos are interesting, clear, and nicely composed. Recordings or videos of your audio work should be a representation of your work as much as the overall show itself is. Remember that everything you post is a reflection on you and your style, creativity, and originality.

Production designer Parker Genoway actively uses social media, and explains his approach:

> Social media is a great way to stay fresh on everyone's mind, even when you are not physically in front of them. Every once in a while, I throw in a picture of a show I worked on. I will get responses such as, "Wow, I did not know you worked on that!" Plus it becomes a sort of resume where people can take a few seconds to scroll through my feed and see my design tastes; then make a decision off that.

When specifically thinking about social media self-promotion within the live entertainment audio field, there are certain procedures that should be followed. Most productions do not want postings of pre-production, backstage, or other candid photos without their permission. Always find out the specific rules for a show regarding content sharing. A good rule of thumb is that you can always re-share something the artist or the management has shared. If they have already made it public, then you can freely re-broadcast the same content.

Remember to always ensure that your photos or videos represent not only your work well, but also the show, the artist, and other designers. Above all, do not leak any secrets or special moments within the show.

Many will wait until a production has ended or a tour has been out for several weeks before posting images. Sometimes it might just be a shot of someone at a console with the show in the background, while others might be videos of the production in progress. Remember to never post before or during a show, and rarely immediately after the show ends. Instead, wait from a few weeks to a month before sharing your involvement. Use official event and artist tags when possible to widen your audience.

You can also make use of platforms and features to post obscure "hints" about what you are working on. Perhaps this means a close-up of a console or the PA in a rig, with a tag of the city where you are working. Be certain not to provide any identifying information about the show. This will build interest in your current project and invite people to follow you so they can learn more as information becomes available.

Some productions will require Non-Disclosure Agreements (NDA) that legally prohibit you from talking about or sharing information regarding a production. Always adhere to these rules to stay in compliance with the production's wishes. You don't want to get fired or sued due to a violation of an NDA.

Always use honesty and clarity regarding your involvement in the sound of a production. Do not just post a photo of a band on stage and state: "I had a great time touring with Band X." Instead, state your involvement and give proper credit to the sound mixer, system engineer, and the rest of the team. For instance, a better posting would state: "It was a pleasure to tour on the audio crew of Band X with FOH engineer Mr Z." Honesty and transparency are principles that should be adhered to regarding all your content.

Since your individual social media represents both you and your work, keep your personal posts about yourself professional, and throw in work-related images/updates from time to time. Allow people to see your personal interests as well as your work abilities.

When properly used, social media can help you engage with others, increase your network, and in some cases, even land a new job or gig. Genoway was able to gain a new client via his postings and audience on social media. Genoway explains:

> I posted a photo where I tagged both the location (the Coachella Valley Music and Arts Festival) and the artist with which I was working. The manager for another artist was scrolling through the official Coachella festival feed and found my post (due to the tag). He then reached out to me because he liked the aesthetic of that particular performance, and he was curious who had been the production designer. After several subsequent conversations, he asked me to submit a design for his artist.

Social media platforms are ever evolving, so be sure to keep up with the latest features, trends, and systems. Stay active and engage with others to promote

Table 11.1 Social Media Guidelines

DO	DON'T
Be friendly Only leave kind feedback and helpful remarks.	**Be a know-it-all** Share information, but keep it within reason. Don't respond if you don't have a helpful answer.
Post regularly At least every two weeks for work-related items.	**Post everything** No one wants or needs to see what you have for dinner every night.
Be honest and transparent Give credit where credit is due and be clear about your role within every show.	**Reveal secrets** Surprise appearances, product reveals, guest artists, and favorite gags should be saved for the show's audience.
Participate Share content from peers, respond via comments, send links to others.	**Set it and forget it** Platforms adapt with the times and so should you.
Connect multiple sites Link together platforms so you can post from one location. Many tools exist to facilitate this process.	**Promote only yourself** Share the work of others that you admire and give them proper credit.
Build your audience Gather many tailored followers who can contribute to your future employment and wellbeing.	**Blatantly ask for work** Posting that you need a job is like begging on the street corner.
Follow production guidelines NDAs, backstage photos, production requests, and secrecy should always be respected.	**Rely solely on social media** Many people avoid social media and if social media is your only promotional outlet, you will miss them.
Understand the algorithms Know what gets featured and is likely to trend. Optimize content for the best distribution.	**Post bad content** Photos and videos should have good composition, focus, and interest. Ensure the audio track of every video is clear.
Update your settings As platforms grow, adjust your setting regarding features, usage, privacy, and outreach.	**Ignore social media** Since social media platforms have a global user base of billions, it would be ridiculous not to participate at some level.

yourself and your accomplishments. Remember that while social media is a valuable self-promotion tool, many people still avoid using it. You will run across many that never see your social media activity. It should only be one tool in your promotion toolbox.

Industry Media

Within the live entertainment audio business there are several media companies that produce magazines, blogs, websites, podcasts, and other sources of industry news. They are each hungry for content and are a great source for your self-promotion. You can reach out to any of them to share the details of a recent production or venue that you have been involved in. You can ask them to create a story around you and your work. Furthermore, you can often write a press release or a case study of your accomplishments and then submit this to them for quick publication.

Prior to reaching out to a publication, always check with the production's management regarding photos and information. Once all the approvals are in place, work with the magazine to ensure that they write an article about the show and all persons and companies involved. This will put your name in front of the industry and help promote your work. A strategy like this is a possibility for anyone in the industry. All it takes is the time and effort to reach out to the press and let them know what is happening in your world.

Another avenue is to contact the audio equipment manufacturers and have them write a story about your work with their products. They will then distribute this information among their customers as well as the industry press. This will often result in quicker publication and distribution as they work with the industry media on a regular basis.

Don't flood the press with everything you are doing, but be diligent about getting your name associated with productions a few times a year. This will further your reputation in the industry and might lead to new opportunities as well.

Word of Mouth

It is often said in the entertainment industry that "it is who you know," and this is certainly true for audio. First, you must be good at what you actually do, and then whom you know will play a great factor in your career. Production managers, shop managers, sound designers, FOH engineers, crew chiefs, and account reps are often asked whom to hire for various jobs. When you introduce yourself to anyone, be sure to give them your full name and look them in the eye. When working on a production, take the time to meet as many people as possible. If you simply keep your head down and mix the sound as instructed, no one may know who you are to ask you back for the next show.

Your reputation will precede you, so be sure to always do your best, as you never know who is watching. If you have a bad attitude and are hard to work with, word will spread, and few jobs will come your way. However, if you are personable and hard-working, you will remain busy.

Take the time to share your contact details with others, as they will likely be in situations where they need your services. All too often, I have seen technicians, mixers, assistant sound designers, and others refrain from providing basic personal information with others. You never know where your next gig will come from, so make yourself known to as many people as possible. When a production or tour ends, take the time to thank each person and share contact details then too. A good follow-up email after the show can go a long way and help you secure future work.

Business Cards

You should also produce business cards that state your name, phone number, email address, and website URL. This is a good thing to share when you connect with someone or when they ask for your contact details. The card should state something about what you do such as "audio technician" or "sound designer." This way, when they look at the card a few days later, they will remember the reference.

Never hand out your card to solicit future work while on a current production. Even if you think you can do a better job than the current engineer or designer, it is not the right place to tell the client that they should hire you next time. This mistake will always lead to the original engineer or designer finding out and you will be seen as a job poacher. Tread lightly when handing out business cards and only give them to the client when requested.

Bring your cards with you to industry events, tradeshows, audio shops, and any other locations where you could meet potential clients. There is nothing worse than having someone ask for your card and you don't have any with you.

Share Your Knowledge

No matter your current position or status, chances are that you have some unique experiences or knowledge that could be valuable (or at least interesting) to others. There are many resources through which you can share your knowledge, and by doing so you will also be promoting yourself. Table 11.2 details some key opportunities.

When employing any of the above tactics, be sure to include references to your website, email address, or other methods to contact you directly.

Figure 11.1 Some Business Card Designs Are Better Than Others

Table 11.2 Sharing the Knowledge

Instructional Videos	YouTube is filled with videos demonstrating everything including sound mixing, system engineering, microphone placement, drawing plots, and other audio principles. Find a skill that you are comfortable teaching and make a video explaining your process.
Blog (create or contribute)	A blog is a website where people post and/or contribute writings about a particular subject. Create your own blog to share your audio knowledge and experience, or participate in an existing blog as a contributor.
Podcast (create or interview)	A podcast is a series of audio (or sometimes video) recordings of discussions and information. The number of entertainment audio-related podcasts is continually growing. You could take part in any of the existing podcasts or create your own.
Learning Seminars (create or take part)	Many audio companies host events for which they invite local industry persons to attend and learn about the company and their offerings. Some of these events also look for individuals to teach sound information. In addition, you could create your own training event, or market yourself as an expert presenter to teach your skills to theaters, production shops, unions, and schools.
Tradeshow Sessions	There are several key tradeshows within the industry such as NAMM and PL&S. They are filled with learning sessions, case studies, and other teaching opportunities. Each tradeshow starts looking for people to moderate sessions many months before the show, and will entertain any idea you wish to present. This is a great opportunity to share your knowledge (and your name) with the industry.
Manufacturer Case Study	The various audio equipment manufacturers each have their own marketing department, social media, and website. Similar to the press, they are always looking for content specific to their products. If you present them with a case study as to how you uniquely used their products, they will share the information on their website and social media while also promoting you at the same time.

Furthermore, use your website and social media to promote these experiences in which you are sharing your knowledge.

Socialize

Computerized communication comes in many forms and can be extremely useful for promoting yourself and your work. However, actually meeting with other people in the same business usually forms better connections. Many in the industry are willing to share their knowledge, tell stories, let you watch them work, provide advice, and more. The real key is to ask! Reach out to others who are doing the type of work you do or wish to do in the future. Meet up at industry events or create your own opportunities.

If you have goals of working on Broadway, then go to New York and ask designers, associates, and technicians to join you for a cup of coffee. Then discuss their careers and explain your goals and desires. Hand them a business card and keep in touch after the meeting. Chances are that they might have an opportunity for you at some point. Even if it is not paid work, it could be the chance to sit in on a sound check or a rehearsal for a Broadway show.

If travel is not possible, then reach out via email and arrange an online meeting using a program such as Zoom, Skype, or Google Meet. Use your video camera and stay focused as if you were sitting with the person in real life. It is amazing how well a virtual meeting can replicate an in-person meeting. A good old-fashioned phone call also works well to stay connected with others.

The more time you spend talking with actual people in person (or via online or phone), the more your name and talents may lead to work. If others know who you are, they are more likely to hire you or suggest you for an opportunity.

Paying for Promotion

Designers and some engineers have found it beneficial to hire an agent to help them secure work. Just as a theatrical agent will find auditions for an actor, a designer's agent will work with producers, production managers, artists, and others to promote you and your work. Of course, this is not a free service, and thus they will take a percentage of the fee they negotiate for your work. Usually it works out well for both parties as you gain additional opportunities with only a slight reduction in pay. In some cases, the agent will ensure you get the rate you want by just asking the client for a higher rate to cover the agent's fee. On the other hand, an agent can also negotiate an even greater rate for you than you were expecting.

A designer's agent is an excellent tool to help keep a designer working and expand their career. It also lifts much of the burden from you to promote yourself, as you will have a cheerleader in your corner booking gigs for you. Some people find that agents can be limiting, however, as they may only work with a specific circle of clients or a genre of production. Look for an agent who has a broad range of experience, clients, and contacts beneficial to you.

Remember that self-promotion is a key part of your job regardless of your position in the live entertainment audio field. It is a full-time job that assures increased work and future career growth. As the popular saying goes, "any press is good press," so take the time to spread the word about *you*.

12 Networking With Others

The live entertainment audio industry is largely based on relationships and thus being skilled at networking with others is very important for your career growth. If you simply do your job and no one on the crew knows your name or how to reach you, then you likely will not get many other calls for work. Whether you are an extrovert or introvert, it is important that you understand the importance of networking.

Make Contact

The real key to success in any career, besides being good at what you do, is having people know who you are and what you are capable of doing. No matter what role you have in the audio field, you must continue to meet and socialize with others to increase your potential. This can be easy for some, but very daunting for others. Just remember that for success, you must step out of your comfort zone and interact with others.

Start locally and talk with people you already know about audio. Find coworkers and peers with whom you can discuss recent shows or interesting articles, concepts, and ideas. From there, branch out to people you have not yet met. Send emails to designers or engineers who live in your city and ask to meet for a chat. Even better, make a phone call to an audio professional and tell them you want to learn more about how they have grown their career. Meeting face-to-face via an online platform helps form stronger relationships too.

Many in the business travel for a living and you might discover that someone important may temporarily be working in your town. If you are following a particular designer, FOH engineer, or crew member, keep track of when their show is coming to your area. Then reach out a few days before and see if they have any time to meet. Often there will be a day off, or even a few hours before doors, so they may appreciate meeting up with someone new. In all cases, never ask for free tickets to their show.

When you are networking, be sure to introduce yourself with your full name. Listen carefully to what the other person says and ask questions about them and their work. Do not directly ask for a job, but instead let them know that you are open to opportunities to grow your resume. Hand them a business card as

DOI: 10.4324/9781003188957-14

Table 12.1 Conversational Skills

DO	DON'T
Consider body language Notice both yours and the other person's to determine openness to the conversation and emotional responses.	**Multitask** Stay present in the conversation and away from your devices or other distractions.
Be confident Trust yourself and your knowledge.	**Interrupt** Let the other person finish speaking before you speak.
Make eye contact Maintain eye contact as much as possible. Scan across multiple people when talking with a group.	**Speak quickly** If your speaking pace is fast, they may not understand, especially those that are from other countries.
Nod Use your head to convey joy, agreement, disappointment, or sorrow.	**Cross your arms** Closed-off body language will discourage others from talking with you.
Focus on the other person Keep the conversation flowing by allowing others to talk about themselves.	**Forget to give your full name** Ensure that others know who you are by name.
Listen Don't just wait to talk; instead, listen and take in what the other person is saying.	**Squirm** Stillness will confirm that you are listening and interested in the current moment.
Ask questions Start with a basic question, then ask the other person to elaborate on their answer.	**Only talk about lighting** Learn more about the other person than just their lighting interest and occupation.
Remember names Use the other person's name when speaking with them.	**Share secrets** If something is supposed to be a secret, keep it that way.
Be genuine Honesty and integrity from you will encourage others to share accordingly.	**Look at the floor** Stay connected to the conversation and the person in front of you.

a means of providing your contact details. Do not provide the card expecting work opportunities in response; remember, it is merely a document with your contact details.

Once you make connections with others, maintain the contact with regular phone calls, video chats, texts, or emails. The best choice is a face-to-face meeting, but this is often difficult to arrange with busy schedules. The frequency of your contact with these individuals should be tempered by how involved they are in the relationship. Always use caution and make sure you do not become an annoyance to someone who is not interested in sharing with you.

Look for Opportunities

Networking should not only happen with contacts that you may know or wish to know. Find industry occasions to meet new people. For example, you can attend tradeshows as big as NAMM or as small as a local shop vendor show-case to make new connections. When you arrive at a trade event and visit a vendor's booth, introduce yourself and talk to the staff about your career desires. Frequently you will gain great advice and connections. Too often, peo-ple simply look at products and do not realize that others standing in the booth may just be the key to that next big career step.

There are also other industry events such as training sessions, meetups, and competitions. Try to attend as many as possible so you can meet the other attendees and hosts. Explain what you do and what you hope to do. Listen to their career goals and see where you might be able to work together. During industry cocktail receptions, make it a point to meet new people and talk to many. Remember that it is up to you to put yourself in a position of opportunity as much as possible.

Give to Get

Networking is not a one-way street. When you are talking with peers, mentors, and others, be sure to offer how you might be able to help them. Maybe you can volunteer to bring them coffee during their next rehearsal or take notes dur-ing tech. Perhaps you can help them organize their workbox in the shop. Find a way to offer assistance to the other person.

Remember that regardless of your level in the field, you have some informa-tion to share. There are also many ways that you can offer your skills to help another who might not have the time or knowledge to achieve what you can provide. Ideally, networking should provide value for both parties. For instance, you might mention to a newly met system engineer that you recently took a course in drafting. This could lead to you drafting for their next production.

Follow Through

After you meet someone, be sure to send an email or text thanking them for the time. Be simple and direct and let the person know you enjoyed the interaction. Then, reach out to this person from time to time (without being annoying) to keep the interactions going.

As you develop your networking skills, you will find it easier to call others and reach out to people you do not know. It takes time, perseverance, and prac-tice, but the rewards are immeasurable. Luckily, most people are happy to talk about their craft, and from this, connections often form that lead to additional work.

Networking should be a lifelong action that you perform at every possible occasion. With more connections, you will discover more prospects. Once you start your interactions with others, you will find it easier and easier. As with many things in life, the hardest part about networking is getting started.

13 Keep on Moving Forward

The entertainment audio field has many options and paths that one can follow. As you navigate through your career, it will be up to you to follow and grow your own path. Remember that taking little steps leads to big changes. In order to move forward, you must continue to learn, promote yourself, and grow as an individual. Throughout life you will need to work hard on yourself in order to advance in your career.

Your Inner Voice

The audio industry provides many people the opportunity to follow their passion and work on productions, events, and other entertainment genres. Even with deep passion and drive, the human mind often intervenes to set one back. Your inner critic may tell you that your work is not good enough or that you do not have the skills to do larger shows. Perhaps your mind steers you away from taking certain gigs or discourages you from taking creative risks.

Early in my lighting design career, I often questioned the positive remarks I received from collaborators about my design. I was not sure that I could trust their opinions or even their word. For instance, one director would tell the set designer that he liked the set and then complain to others about how awful he thought the set looked. How could I then take any positive reinforcement from him about my lighting with any degree of respect?

From there, I began to worry about any comments that were made about my lighting designs. I spoke with my father about these inner thoughts and he explained that if I had done my best, then I should be proud of my work. He also said I did not need to be concerned about other's opinions; that if I was continuing to be hired, then I must be providing good lighting. I had to learn to silence my inner negative dialogue and only listen to praise and constructive criticism from others.

Passion to Improve

Recently I was teaching a class when a student asked for some advice. He was concerned that he was stuck in his current career position and did not know

DOI: 10.4324/9781003188957-15

how to proceed. He said that he felt intimidated by "other people that kn[e]w more than him." He was scared to take bigger gigs or high-responsibility positions as he was comparing himself to others. When I asked him specifically what he was apprehensive about, he told me that he did not feel he was as skilled as others he had met.

I explained that there would always be people with more experience and knowledge than him. He needed to accept his current stage and sell himself at that skill level. I suggested being open with clients about his knowledge and not taking on opportunities above his existing skillset. Next, he needed to prioritize improving his own skills over almost everything else in his life. If he truly wanted to improve as an engineer, then *he* had to make the decision and make it happen.

I suggested that he focus all his spare time on improving his engineering knowledge and skills. This meant turning off the television, ignoring social media, and not going to the bar. Instead, he needed to spend his time reading the user manuals, practicing on key software, and going to shops or venues to borrow time on real equipment. In order to fight the intimidation he was feeling, he needed to turn that energy into excitement to improve.

People with a passion do not need to be told to work harder or learn more. If you have a deep passion for working in the entertainment audio industry, then you should apply copious amounts of your time, thoughts, and effort into doing your best and growing on multiple levels. This should also include working on yourself and your own mindset.

It seems that our minds are often our worst critics; thus, we need to learn to direct our influences away from ourselves. This can be achieved by coming to terms with one's own thoughts. Meditation, personal growth, and therapy are efficient tools for controlling our thoughts. Many books, videos, courses, blogs, spiritual advisors, and podcasts are available to help you dig deeper into your mind to focus on yourself and your passion.

Be Open to Change

Maybe you desire to become the next great sound designer. However, along the way, you find that you just don't have the interest that you once had, or perhaps that you lack some of the creative or communication skills required of that position. From time to time during your career, you might need to reevaluate your goals and decide to change your path. Or perhaps opportunities will arise that take you in another direction altogether. For instance, you may discover that you are better at being a FOH mixer for television productions than as a theatrical sound designer. By following this alternate path, you could achieve a wonderful career working on amazing television productions.

Sometimes change is not easy and may not be of our own choice. I know one sound designer who had a rough couple of years with very little work. She had to work at various "normal" desk jobs to pay her bills. Then one day she answered the phone and was offered a job as a production manager at a large house of worship. Although the position did not fit within her ideal career plan, it was a paying position within the production community. She soon found herself managing a staff of people including audio and other technical crew. She was able to apply her organizational skills learned from being a sound designer to her new position and carve out a great new career path.

Whenever you are accepting a new job, position, or gig, remember that you should strive to be an active architect of your own work. While many positions are clearly defined, the actual duties involved may or may not be exactly what you desire. Take the time to discuss with your clients, manager, or supervisor the expectations of your position and tell them what *you* want from the job. If you are hired as an audio technician on a tour, you may not be able to ask for many things, but you can be clear about your skills and abilities. If you desire to become a mixer or designer, then ask for the A1 position (assuming you have the skills to do so). Otherwise you might get relegated to working backstage and miss opportunities to learn about mixing and design.

Sometimes, a more senior person will see a potential within you and decide to provide you with a new opportunity. I know one touring audio technician who was offered training and a position to operate the truss automation for a tour. Although it was not an audio position, the production manager knew that this technician was interested in many different areas of production, and that he had the discipline required to operate motors over a stage. The audio technician learned the automation system and went on to program and operate the automation for a tour. Although he enjoyed the automation gig, he confirmed that he preferred audio and reverted back to audio for subsequent tours. While this deviation from audio did not result in a career path change, it did help him acquire more skills and contacts that he has since been able to apply to his audio career.

People often resist change, but you should look at change as a chance for growth. Even negative change such as being fired or laid off usually leads to better things in the future. If a job is not a perfect fit, change it to better match your desires within the scope provided by the employer. Remember that only you are in charge of your own career, so make it the way you desire.

Defining Success

The definition of a successful career is a personal one. You may think success means you must be a mixer for major bands, a designer for Broadway shows, or winning certain awards. But there are no set requirements for being successful in the entertainment audio field. Success should be defined as: "whatever makes you happy." For some this is a perfect work-life balance, for others a certain amount of money. Some may assign goals pertaining to types of shows or even winning awards.

I have seen very successful people who built great lives for themselves and their families while enjoying their work. To me this is the best marker of success. Personally, I have achieved some amazing things and still plan to accomplish more. There are others who have worked more high-profile events or earned more money, but that is not how I define my success. Decide early on your markers for success, but allow them to change throughout your life as events unfold. For instance, you may wish to mix for a popular band, but later realize that the same mix engineer has been doing it for over 30 years and is not looking to retire.

When deciding on your own personal success, remember to only compare yourself to yourself and keep your dreams within reality. Realize that results do not happen instantly and allow time for your success to manifest within your career. Hopefully, with hard work and perseverance, you can look back at various points along your path and see the successes and growth within. Then you will truly be able to feel triumphant regarding your career choices.

14 Giving Back

The live entertainment audio industry is comprised of a rather tight-knit group of people, with many looking for opportunities to share experience, knowledge, and skills with others. When you are first starting out, you might think that you do not have much to offer, but this could not be further from the truth. There is always someone out there who can learn from you or benefit in some way from your assistance. As you advance in your career, the opportunities to share will expand exponentially.

Teaching

The most common method that people think of to "give back" is to teach or share their acquired knowledge. This can happen in numerous different ways. One might explain to another how to operate a sound console or perhaps create a "understanding speaker placement" video to share online. See Table 11.2 for suggested teaching ideas.

As you teach, make sure your audience understands that you are sharing your skills and experience. When they ask questions, listen carefully. If you do not know the answer, then let them know that you will research it and get back to them. Write down the questions, then after the training, take time to learn the answers. This then becomes your reward, as you will learn something that you may never have even thought to ask. Remember to report back your findings to the person who originally asked the question.

Industry Charities

Around the world, there are several different charities that have been set up to assist people working in the entertainment industry. In North America, the one most commonly known in the entertainment technology industry is Behind the Scenes (www.behindthescenescharity.org). They provide financial and mental support to entertainment technology professionals (not just sound) in times of need. Grants are customized for each individual or even their immediate dependent family to assist with living, medical, and/or funeral costs. They also have a

DOI: 10.4324/9781003188957-16

special fund that allows access to mental health and addiction counseling. In the United Kingdom, a similar group exists called Backup (www.backuptech.uk).

As a member of the industry, you should be aware of these charities in the event that you or anyone you know ever needs their services. Furthermore, you can help the charity of your choice by making financial contributions. Each will have ways to make direct donations, but there are also some other ways to contribute.

One method is to pledge a portion of your income to go directly to your charity. This will help the charity immensely, but also may be a tax-deductible item to help you reduce your tax burden. Another common way to donate is to link a charity to your Amazon account via smile.amazon.com. Once registered, Amazon will give a portion (currently five percent) of the money you spend on purchases directly to your charity of choice. This option literally costs you nothing as you were already going to give the same dollar amount to Amazon. Keep in mind that, in this case, Amazon gets the tax deduction instead of you.

Local Charities

Beyond the large industry charities, many local options are available. For instance, in Nashville, lighting designers Chris Lisle and Erik Parker saw the need for a local resource for touring staff to learn life skills. In 2011, they created a non-profit charity called Touring Career Workshop (www.touringcareer workshop.com). They host an annual free event aimed at touring professionals that provides all in attendance with information about business, taxes, insurance, retirement, relationships, and more. Sessions are presented by professionals who also donate their time by sharing their personal experience and knowledge. The Touring Career Workshop has grown over the years and now includes several events each year as well as free counseling, tax advice, life coaching, and more through their All Access Program.

This is a great example of two industry veterans who created a way of giving back that directly helps many in the industry. From online portals such as gofundme.com to regional campaigns, many avenues are available for anyone to promote a cause and help others in need. Many unions also have charitable causes benefiting their members.

Remember that charity does not just mean donating money; it could be sharing of your time and knowledge through various resources. For example, you might offer to help a local high school with the sound for their annual musical or arrange a training class after hours at the audio shop where you work. Charitable opportunities are abundant if you take the time to look for or create them.

More Giving

Most charities will hold events such as raffles at tradeshows, golf tournaments, and competitions to help raise additional funds. Your participation in these events (either as a contributor, sponsor, or a donor) will be greatly appreciated. Probably the very best option with any charity (other than donating money) is to help those in need become aware of the services available to them. Share with your colleagues, talk about the charity on social media, and post signs in your shop or venue. Basically, do what you can to increase the awareness to help all in our small industry.

Many designers and technicians will also take part in charity events that are unrelated to the industry. Perhaps you decide to donate your time for three nights mixing the sound for a cancer fundraising event. Instead of charging the producer of the event for your time, you simply complete the work at no charge. In this manner, you are using your skills to assist the event and reducing their operating costs. In turn, this allows them to raise more funds. Remember too that you can then apply the amount you would have normally charged for those three nights as a charitable donation and possibly reduce your tax burden at the end of the year.

Mentoring

A favorite choice for many in giving back to the industry is mentoring someone who is starting out. Many sound designers or mix engineers will allow students and up-and-coming professionals to come and sit in the theater or at FOH during the pre-production of a show. Sitting quietly a few feet behind sound designer Nevin Steinberg or FOH Mixer Vincent Casamatta, listening to how they work with their equipment and team, is an amazing and powerfully educational experience. No matter where you are in your career, you can always find someone with less knowledge and experience than you. Invite them to watch you work and take some time to share when possible by explaining your actions.

Mentoring can occur onsite at a production or remotely when not working a show. Perhaps you take the time to review the drawings of colleague or offer mixing advice to a friend. Always be open to sharing information, as others have probably been open to helping you.

Always Ready to Share

Even if you are early in your career, you will always have some skills you can share with others. You might explain to another technician your approach for labeling microphones or decide to host a mini seminar in the local production company's office. Just teaching others will help you to solidify your personal skillset. Furthermore, you could post tips on Internet forums and social media or even tweet a new sound trick you recently learned.

I recall a group of Los Angeles-based mix engineers who would get together regularly at a local bar. They would discuss their recent gigs as well as describe any new console functions or tools they had discovered. By sharing this information with each other, they each continued to develop their skills at a rapid pace. Furthermore, they solidified their working relationships and made new contacts for the future.

Beyond sharing knowledge, you can also pass employment opportunities on to others. When you are not able to say yes to a production, think of someone you can suggest for the job. In doing so, you will pass the gig on to another person and everybody wins. This is especially important as you focus your career towards specialization. For instance, when you move from technician to engineer or designer, pass technician jobs to others who are at that point in their careers.

Share the Passion

The audio industry thrives on individuals that are filled with passion for sound. There are always new people coming into the field with strong desires to learn as much as possible. You should always be open and ready to share your skills and knowledge with everyone. At a corporate event I worked, a motivational speaker explained that the word "passion" could be broken down to "Pass I On." She explained that if we are passionate about something, then we should share that part of ourselves with others.

The best way to do this is to share your audio passion through your knowledge and experience. So I challenge you to go out today and share a bit of your sound knowledge with another person and spread your passion. Also look for ways to support industry charities and other methods to give back all that has been given to you.

15 Moving On

All good things must come to an end. This proverb has been said in many situations for thousands of years. Alas, it is true too of your career in the audio industry. Not all people will end their careers; some will just live a life of sound until the very end. Others will choose to change careers, retire, or move into new and different roles. While planning for your long-term future is often suggested, it is not required. However, it is good to know about the choices available to audio professionals.

A Lifetime of Sound

I have found that most people work in the live entertainment field for about 30 years before they begin thinking of slowing down. Generally, after 30 to 40 years of sound, most are ready to retire or at least reduce their workload. On the other hand, a good number wish to continue working at their current pace as long as possible. It is remarkable that the industry is such a magical one that people desire to work within it for their entire lifetimes

Varying Roles

After many years working in a specific portion of the audio industry, you may choose to change your area of specialty. Many seasoned professionals shift to consultant, teaching, venue staff, or sales roles. They will no longer directly work on productions, but rather focus their efforts on a new segment of the industry. There are many options and it can be exciting to change your daily workflow to a new, but related, part of the industry.

Opportunities abound and it is easy to transition to a new role. Often this can be accomplished while still working in your existing position. If you desire to teach or work in sales, you will likely be able to work a few shows too. Plus, as you move from one position to another, you will bring your previous experience and contacts along with you. These invaluable resources will prove extremely beneficial both to you and your new employer.

DOI: 10.4324/9781003188957-17

Changing Careers

Sometimes people will find opportunities outside of sound and change careers accordingly. They frequently find that their experience and knowledge from the audio industry provide unique skills and insight that can be used effectively in their new field. Some opt to remain with production but focus on other areas, like management, video, lighting, or special effects. In some cases, people pivot to a totally different industry. I know one audio engineer who owns a bar and plans to focus on the bar for his future after audio.

The small, family-like qualities of the relationships gained within the audio industry usually help those who choose to move on to other fields. They find that audio associates are always willing to help out and will be there should they return to sound in the future.

Happy Retirement

Many sound professionals choose to save money or make investments throughout their lives to allow them the freedom to not work later in life. This is a great strategy because we all age and probably won't find working as enjoyable or easy when we are in our 80s and 90s.

The exact definition of retirement is deeply personal and varies widely between individuals. Some just want to live comfortably and spend time with family. Others choose to sail a boat, travel the world, or relax in the countryside. Still others will build or invest in a business that they can oversee without being involved every day.

If you desire to have a point in your life from which you do not need to be concerned with having an income, then it is important to start saving for retirement as early as possible. Because many in the industry work as self-employed, you may not have access to a retirement plan or government savings such as Social Security available to you in the amount you require at retirement. Unfortunately, I know several older people who are forced to continue to work in order to maintain their desired lifestyles. Had they saved more money earlier in life, they would now be in a better place to work less (or not at all).

Retirement from sound can be achieved. Most designers, engineers, salespeople, technicians, and others have a great time later in life without financial concerns. Participate in union and company retirement plans, save money, make smart investments, and support your government retirement options as well. With good planning, you can comfortably retire and remain happy for the rest of your life when *you* are ready.

Part 3
The Business

Regardless of your position or stature in the entertainment audio field, you will need to understand basic, and perhaps more advanced, business processes. While some are general employees, many audio professionals find that they become their own small business and must undertake every aspect of running a company. It is important to understand the options and requirements as you move through different aspects of your career.

DOI: 10.4324/9781003188957-18

16 Types of Workers

The Freelancer

Many in the live entertainment audio field choose to work as freelancers instead of employees of a company. There are of course pros and cons to each, and you might even do a little of both during your career. In some cases, you might be an employee at a company and a freelancer on the side. You could even become a "permalancer," where you are a freelancer who continues to work only with one client.

The decision to work as a freelancer should not be taken lightly as there are many things to consider when working for yourself. Some of the advantages of working as a freelancer include:

Make your own schedule: You can choose which productions you wish to work on. If you don't want to work over major holidays or certain dates, then you simply do not accept any gigs during that timeframe.

Choose your shows: Much like being in control of your schedule, you can also choose to decline certain productions. Perhaps you do not agree with the political agenda of a corporate client or you just can't stand rap music. Either way, you can turn down the gigs with which you don't want to be involved.

Figure 16.1 Types of Clients

DOI: 10.4324/9781003188957-19

Easier work-life balance: When you are not working a show, you can focus on your family, friends, and/or pets. Not having to work in an office or meet the workload demands of an organization can be freeing.

Live where you want: Since you are not tied to a company's office, you can choose to live where you want. You may select a city based on the productions in a town, or if you are mostly touring, live anyplace you wish. Many engineers and touring audio staff will reside in a location with low taxes and low costs of living. Because they almost always travel to their gigs, their home location is not tied to anything work-related.

You are the boss: The buck stops with you. There is no need to ask for permission, fill out annual reviews, follow corporate policy, turn in reports, etc. As a freelancer, you don't have to answer to anyone, except your clients.

Of course, there are some significant disadvantages to working as a freelancer too:

Inconsistent work: When it rains, it pours, but wow is it dry when there is no rain! Your workload may be very busy at times, then very slow at other times. During the slow times you will wonder when you will get the next job as you watch your bank account slowly drain. Often multiple gigs will be available at the same time, while you can only choose one. Always save extra money for dry spells.

You are the company: As a freelancer, you are responsible for all aspects of your business. You must handle bookkeeping, taxes, invoicing, marketing, training, and more. This burden can become tedious and take you away from the audio work you love.

No benefits: If you work for a company, they will often provide many benefits that a freelancer misses out on. Insurance, tax assistance, retirement, tax withholdings, unemployment benefits, and other perks will not be available to the freelancer. Always save a portion of your income to pay your taxes. Keep an emergency fund of at least six months to cover any unemployment periods.

Irregular income: Since you get paid per show, your income is going to vary from gig to gig. The amount and regularity of payments to your account is going to fluctuate wildly. You need to prepare for the dry spells with significant savings and manage clients to ensure you are paid. Some productions may take 30 days or more from the end of the show to send a check.

You have to find the work: As a freelancer, you will hope that your phone will ring and the next job will be provided. But that does not always happen. You need to spend a good percent of your time looking for work and promoting yourself.

Higher taxation: Depending on your location and the local tax laws, you may owe more when you work as a freelancer. In the United States, an employer usually deducts payments for Medicare and Social Security. As a freelancer, you are responsible for paying into these systems yourself in addition to paying other self-employed taxes.

More complex taxation: As a freelancer, you will likely have multiple tax forms sent to you by your clients. In the United States, these are commonly called 1099s and they show the government how much each client paid you. You must keep track of this paperwork as well as determine how to properly file your taxes each year.

No coworkers: When you are not working on a show, you will find that you are alone at your home office. Sure, you can call or meet up with others, but there is no one else to help you with various burdens. You can't pass tasks off to others and you certainly will miss out on the camaraderie of company life.

Freelancing is a great form of employment and a large percentage of the industry makes a living in this manner. You need to realize that you must prepare for the freelance way of life by ensuring you always have at least six months of expenses in savings. In addition, you must actively work to promote yourself and manage your business while you are engaged in the audio work you love. With diligence, anyone can have a successful freelance career in the live entertainment audio industry.

Figure 16.2 Types of Companies

The Company Employee

The alternative to working as a freelancer is to work as an employee at a company. Audio professionals have many employer options available including rental/production shops, venues, cruise ships, manufacturers, television studios, corporations, houses of worship, event companies, and much more. Generally, any position in the field can find a place within a company or venue. In some cases, you might not only be an employee but also an owner of the company.

As with freelancing, there are several pros and cons to working as an employee at a company. The positive aspects include the following:

Consistent income: If working as a full-time employee, you should have a reasonable idea of how much your paycheck will be for each pay period. Even part-time employees usually have an inkling as to how much they will be paid on a regular basis.

Benefits: Many companies will provide additional benefits to their employees beyond the employment. Health and life insurance, retirement savings programs, corporate discounts, tax assistance, bonuses, and employee wellness programs are common benefits. Some employers will also provide additional dollar matching to increase your retirement savings. Should you lose your job, you may be eligible for severance and/or unemployment funds.

Resources: There will be many other employees and departments that can help you with your tasks and needs. For example, the accounting department can chase down delinquent clients while the human resources team may assist you with personal matters.

Ladder to climb: Working for a company might provide you the opportunity to move up through various positions. Many people wish to increase their income and responsibilities by changing their duties within an organization. Some very driven people have moved from working in the shop to becoming a salesperson and eventually a vice president, all inside the same audio company.

Job security: Generally speaking, you don't have to worry about whether the next show or production will occur. When you work for an organization, it will have a drive to increase its business and hopefully provide you with a strong workload. If you are employed at a sound company or venue and there are no current shows, then perhaps you can help fix the gear or organize the shop until the next show.

Payroll: A good company will help you save by taking taxes out of your checks. This helps to reduce your responsibility during tax season. For example, in the United States, if you only work for one employer, then you will just get a single W2 to file at the end of the year. Assuming you set your withholdings correctly, you likely won't owe the government more money at tax time.

Company backing: As an employee you may be protected in the event of an accident, lawsuit, or other unforeseen occurrence. Most companies possess workman's compensation, liability, travel, and other types of insurance to protect the company and their employees. Plus larger organizations often have legal teams for bigger matters.

Along with the positive aspects of working for a company come the following negative points:

Company policies: You must adhere to the policies and procedures that are in place to protect and aid the company. They may or may not conform to your opinions and desires. As an employee you must comply with all guidelines. From travel policies to corporate regulations, any directives can be assigned to you.

Less freedom: When you commit to work for a company, you agree to provide a certain amount of work in return for payment. They will expect you to work required hours and days (and possibly even more hours). When you want some time off, you must ask for permission and then only take the time allotted.

You can be terminated: At any time, the company can decide they do not need your services anymore. Or the company might just go out of business. Nothing is guaranteed and no one is safe from termination.

Lower pay: Some companies will overwork employees without compensating for long hours of hard work. They may charge a client a much higher rate for your work to cover the overhead of the company. A freelancer on the same show might actually make a higher income performing the same tasks. You are often trading the company benefits and security for a lower wage.

Employees of companies can receive great benefits and job security, but company life is not for everyone. The decision to work for one should not be taken lightly.

The Union Worker

In the United States and other parts of the world, some craftspeople are organized in unions. A labor union is a group of workers who unite to make agreements about decisions affecting their work. In some cases, the members work as employees under a union contract, while others are freelancers following union agreements and guidelines.

In the audio industry, those who hold design positions are typically in different unions than stagehands or technicians. In either case, you could be employed by a show, venue, or shop that requires union membership. Alternately, you might be a freelancer who the union calls to provide work opportunities, or a freelance union member who follows union standards and policies regarding your employment.

Some tend to look down upon unions, specifically because of their apparent aggressiveness towards work duty separations. However, joining a union can have many benefits:

Wages are regulated: The union will negotiate the rate for your work. They also set a minimum amount that you should receive for your services. This ensures that you are always making a pre-defined minimum amount per show.

Working hours are defined: The union will also negotiate for work periods, breaks, minimum turnaround times, and more. They can set forth a plan that compensates you for long hours worked multiple days in a row.

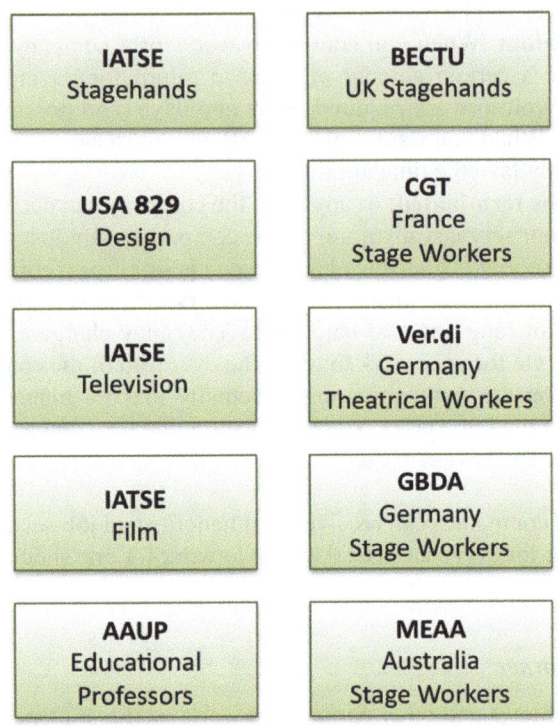

Figure 16.3 Types of Unions

Benefits are available: Most unions offer their members group insurance and retirement plans. This is a great way to help their members and a huge benefit for anyone freelancing. Additionally, unions provide compensation and liability insurance along with programs and counseling for their members. Unemployment funds may be available to union members from local governments.

Workplace safety: Poor and unsafe working conditions are never acceptable. The union will work with the production to ensure safety for all. If a problem arises, the union members stand together to encourage a resolution.

Payment assurance: The union will always work to guarantee that their members are paid by the production. This adds a level of assurance to anyone working on shows, as payment is certain.

Per diem and overtime: Contracted rates also detail requirements for union members regarding overtime payments and per diem minimums. All clients working with union members must follow these and other guidelines.

Training opportunities: Many union offices provide training opportunities to aid their members in career advancement. In-house and external training is often provided at no additional cost to members.

Union employment does not come without its cons:

Regular dues: You will be required to pay the union to maintain membership. Depending upon the union and location, this could be a significant amount. Failure to pay dues could lead to fines and eventual termination of membership.

Union exams and fees: In order to "get into" the union, you will likely have to pass an entrance exam and possibly pay an initial fee. The requirements vary from city to city and also depend on the type of union and skill.

Training requirements: Many locals will not allow their members to work on productions or take specific roles until they complete union-required training. In some cases, you might have to wait long periods until a class is offered or space is available for the training. This could delay your plans to start working or increase your pay potential.

Limited opportunities: Unions tend to book employees based upon seniority. This means the longer you are in the union (or have a family legacy), the more work you could be offered. Those with less seniority may find getting work through the union slow for quite a while. Furthermore, union workers may be limited to certain venues or productions and not have access to all productions happening in a city, venue, or location.

Fee sharing: Designers often have an agreement that a percentage of their design fees will go to the union in addition to the annual dues. In exchange, certain protections, agreements, and benefits are put in place for the designer while working on a production.

Membership may be required: Depending upon your job and location, union membership may not be a choice. For instance, if you plan to work on Broadway productions, you will need to join a union. Many television productions also require membership for designers and technicians alike.

Union workers are just as valuable as anyone else on a production. Unions provide security and support to their members and can be an asset to anyone's career. Union membership is not for everyone and is certainly not a requirement to be successful in the live entertainment audio industry.

17 Business Structure

If you are not working as an employee of a company, you will likely want to set up your own business. All freelancers are essentially already running their own businesses, but most are not legally structured as such. Although you could continue to work as an independent contractor, you would then miss out on business tax reductions, liability protection, benefits, and more. This is why many freelance workers in the industry choose to create a registered business that they can operate under.

Form an Entity

In the United States, there are several different legal entities that can be created, each with its own benefits and downfalls. Some freelancers form a Sole Proprietorship (SP), while others create a Limited Liability Company (LLC) or Corporation (Inc.). In order to set up any of these entities, you will need to follow federal, state, and local laws, which may change over time. In other countries, there are similar types of business structures and specific laws regarding business ownership and requirements. It is always best to get legal advice from a lawyer, accountant, and/or business manager within your country and state or providence of residence.

In addition to varying tax benefits for companies, one of the key reasons to establish a business is to gain protection for the business owner's personal assets from debt or legal liabilities. If you are working as a self-employed person, a sole proprietor, or in a partnership, and subsequently sued for any business reason, a claimant could take your personal assets to recover a court's judgment. With an LLC or corporation only the assets of the business become available to a judgment. You certainly do not want to lose your home or personal savings due to a legal business matter.

Other benefits of setting up a company include availability of business loans, ease of hiring employees, tax deductions, and tax benefits. Furthermore, it might benefit your personal life as well because you can demonstrate that a company regularly employs you. This is usually more favorable than being a freelancer with varying income sources and amounts. For instance, continuous employment is often preferred when applying for a home mortgage, housing rental, car insurance, or personal loans.

DOI: 10.4324/9781003188957-20

Additionally, if you find that you are suddenly without income, it can often be easier to apply for unemployment from your own business than to apply as a freelancer or a self-employed person. Plus, your business might have access to small business loans, government programs, and lines of credit. During the COVID-19 pandemic, businesses were offered special government loans with debt forgiveness that aided many in the industry during the pause of entertainment. Freelancers who were working for their own companies were eligible for this money while standard freelancers were ineligible.

There are many online businesses to assist you with setting up your own company, such as LegalZoom, QuickBooks, and IncFile. Their online forms make registering a business with your state quick and easy, and the service generally costs only a few hundred dollars.

Once your business is legally registered, you can apply for an Employer ID Number (EIN) with the IRS. This free service can be accomplished at www.irs.org. You then provide your EIN to clients in place of your personal social security number. This will demonstrate to the IRS that fees were paid to your company as opposed to you as an individual. Countries other than the United States have similar systems for categorizing business tax information vs. personal tax information.

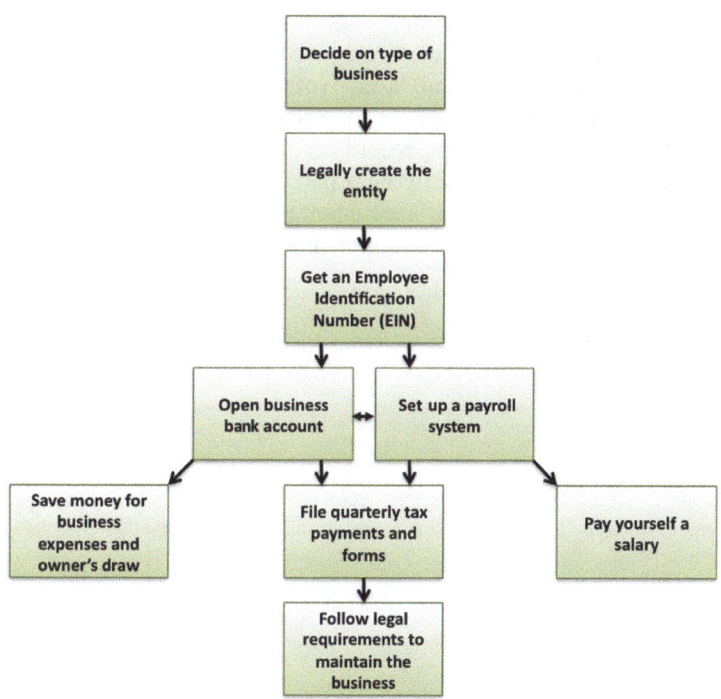

Figure 17.1 Business Processes

Managing Money

After creating a legal company, you should open a business bank account (you should do this regardless of your entity type). Keeping your business income and spending separate from personal finances is a good idea for anyone, but essential for a legal company. This ensures that the business dollars remain in one place, and it is easy for you (or any lawsuit judgments) to track the assets of the company. Remember that the money earned by your business is the property of the company and not your personal income. The company will need to pay you for your services, while hopefully still maintaining a profit for the owner(s) (you of course are also the owner).

You will then pay yourself a salary or take the owner's draw (profit) from the company account as needed. Just remember that you may need to pay additional income tax on these withdrawals, so be sure to keep track of everything. Online services such as Gusto, QuickBooks, and ADP provide simple-to-use automated payroll resources.

Taxes and Laws

Once you are the owner of a small business in the United States, the IRS will require you to submit quarterly estimated tax payments. Form 1040-ES and other documents (depending upon your business type) may also be required quarterly. Additionally your country and/or state may require various payments or documents, so be sure to check your local and federal laws.

Speaking of state and federal laws—your business entity may need to adhere to certain laws and procedures in order to maintain its business status. Required bylaws, board meetings, record keeping, unemployment fillings, and other documentation will suddenly be a part of your responsibilities. Again, consult a business lawyer and/or accountant to confirm that you are meeting the various legal requirements for your business.

18 Personal Business Paperwork

Whether you are working as a freelancer or directly interacting with clients on behalf of a company, it is important that you understand common business documents and essential daily business routines. These include contracts, timesheets, invoices, and other deliverables that your client will expect from you.

Contracts

The term contract refers to a written or spoken agreement that is intended to be enforceable by law. When explaining employment, the contract should describe the scope of the work and the compensation for said work. In some cases, a company hiring you will send you their standard contract to sign. Other times you will need to send them your own contract or letter of agreement for their representative to sign. In additional situations, a simple email, text, or phone call may substitute for a formal contractual document. In all cases, it is best to get agreements in writing. When negotiating the terms of a contract, try to always be reasonable and don't make outlandish demands.

Regardless of the format, the contract or agreement should contain the following key elements:

- **Introductory Statement**—Your name, the client name, the project name, and the dates of enforcement.
- **Terms and Conditions**—Describe the work you will provide in general terms and the amount the client will pay for that work. Also, the conditions for payment (and any deposit) should be listed.

  - You should include details about travel expenses, per diem, overtime, and other clauses concerning your scope of work and the client's obligations to you.

- **Legal Disclaimer**—You may want to describe specific legal limitations regarding your work. For instance, you might have a disclaimer on the topic of the actual rigging of a speaker plot based on your design. The goal is to provide information to the client that indemnifies you from a broad

DOI: 10.4324/9781003188957-21

range of consequences that a client may believe they suffered as a result of your working for them.

- **Copyright**—Depending on your position and your deliverables, you may need to claim or release ownership of intellectual property. This is especially important when providing sound design elements such as effects and music.
- **Payment Terms**—Define when the client will pay you and at what regularity. For example, if you are working on a tour, this section should define how much you are paid per week and list the pay schedule and methodology. For other events, this section should include the number of days that may pass after the receipt of an invoice before payment is expected (usually 30, 60, or 90 days).
- **Signatures**—The document should be signed and dated by you and a representative of the client. Both parties should have a copy of the contract or agreement that includes all required signatures.

It is a good idea for you to have your own contract or letter of agreement ready to send to clients when you are agreeing to work for them. Many sound/production companies will simply sign your agreement and move forward with the booking. Others may send you their own contract and ask that you instead agree to their terms. Some productions may also use standard union contracts and ask you to comply accordingly. Remember that a contract can always be negotiated and that you should always ask for what you think is fair.

If you are a union member, be sure to follow union standards within all agreements and always bring negotiation concerns to the proper union representatives. Designer-specific contracts and agreements from USA829 (United Scenic Artists) are available at www.usa829.org/Contracts.

Figure 18.1 is an example of the letter of agreement I have used when programming lighting for various productions. A quick online search for "freelance contract template" will provide a host of samples and ideas to help you formulate your own contract or letter of agreement. Remember that a contract is used not only to agree on payment but also to set the scope of work and the associated boundaries.

Depending on your role, you may need to define further matters in your contract. Designers may specify first right of refusal for future productions/tours, access to artists, or copyrights regarding their design, music, and cueing. Mixers often include a clause indemnifying them from the responsibility for the rights and ownership of any audio content that the production provides.

As mentioned previously, sometimes it is acceptable to work without a specific contract. If you are working as a freelancer for a production company, they may just call or email you an offer for a show or tour. If you simply reply "yes," then this can constitute a legal agreement. However, it is always a good idea to clearly state in a reply the job you will do and the amount of money you expect in return. This way, if there are differences of opinion later on, you have email or text communication to refer back to. When an agreement is made verbally, always send a follow-up email restating the terms that you have verbally agreed to.

Letter of Agreement

Homer Audio   **Brad Schiller, Inc.**
742 Evergreen Terrace   **123 Fake Street**
Springfield, State 55555   **Austin, TX 78758**
614-555-1212   **512-555-1212**

Dear Homer,

Our signatures on the bottom of this letter of agreement will signify our acceptance of the terms described herein regarding audio mixing services for Homer Audio during a period of eight days on April 21-28, 2022. In exchange for services provided, the following items for payment and reimbursement are agreed to:

- The day rate for services is $xxx.00 for a 10-hour day. Any hours beyond the initial 10 will be charged at a rate of $xxx.00 per hour.
- Travel days and on-site days off are $xxx each.
- Each on-site day (work, travel, or day off) also requires per diem of $65.00 to cover meals and related expenses.
- All travel expenses (car rental & fuel, hotel, taxis, etc) are to be paid by Homer Audio in advance or reimbursed upon receipt of the invoice.
- Travel arrangements are to be made by Brad Schiller, Inc. and reimbursed by Homer Audio.
- Cancellation of a booking within 30 days of the event will result in a charge of 50% the day rate for the booked period (signing this agreement constitutes a booking).
- Payment is due upon completion of the event or within 30 business days of receipt of the invoice.

Brad Schiller (signature)
Brad Schiller   Homer
03/19/2022

Please sign above and email to the following:
brad@bradschiller.com

Figure 18.1 A Sample Letter of Agreement

Time Sheets

Many audio professionals work on jobs paid at a specified day rate with provisions for overtime. For instance, you may agree to a daily rate for a ten-hour day (plus meal breaks). Any hours above this will be paid at a higher hourly rate, usually at time-and-a-half. This is very common for corporate events, television productions, and some special events. It usually becomes your word against the client as to how many hours of overtime you worked. If you do not track overtime as it occurs, it will be difficult to remember after the show the exact number of overage hours you worked. In order to simplify this, keep a timesheet of your work hours to enable you to easily refer back at the time of billing. Many apps such as Clockify, Toggl, and Harvest further simplify the process of keeping track of working hours.

I prefer to use a simple Excel spreadsheet and manually enter in my hours per production. When I am provided with the initial production schedule, I will start a spreadsheet that lists the planned hours for each day. Then at the end of each working day, I will place the actual hours worked next to the scheduled time. Then in a third column I will list the number of overtime hours. Per my contract, my working time starts from the call time and ends at the time I leave the venue. In many cases, I will amend this to the time I leave for the venue and the time I return to the hotel. Travel time does factor into your working time if the client chooses to put you at a hotel far from the venue.

You must be disciplined at maintaining your timesheet daily and ensuring it is accurate and not padded. You can also track time you spend working on the production prior to arriving on site (such as time spent on designing or drafting). If payment for this time is a provision in your contract, you surely want to charge the client accordingly. While I utilize a spreadsheet, there are specialized apps and online services that can also be utilized for tracking your working hours.

Simpson's Sales Meeting				
	Schedule	Actual	OT	Notes
Monday June 20	Travel Day			
Tuesday June 21	8am-7pm	7:30am-6:30pm	0	meals on own
Wednesday June 22	8am-7pm	7:40am-7:20pm	1	meals on own
Thursday June 23	8am-11pm	7:40am-11:30pm	4	Dinner provided
Friday June 24	7am-11:30pm	6:40am-11:40pm	5	ALL meals provided
Saturday June 25	8am-9pm	7:40am-10:20pm	4	ALL meals provided
Sunday June 26	7:30am-10pm	7:10am- 9:20pm	3	ALL meals provided
Monday June 27	7:30am-11pm	7:10am- 12:20am	6	ALL meals provided
Tuesday June 28	7:30am-10pm	7:10am- 8:40pm	4	ALL meals provided
Wednesday June 29	Travel Day			
		Total OT:	27	

Figure 18.2 Example of a Timesheet

Invoices

A very important document that you must become familiar with is the invoice. Most companies will require you to submit an invoice (especially if you are working as an independent contractor). The invoice is simply a document describing the work completed, the amount due, and the terms of payment. In short, it is a bill that you provide to the client. Your invoice does not need to be a fancy document with creative graphics; instead it should be clear and to the point as to what amount is owed and why.

Much like a contract, there are many formats of invoices to choose from, but they all share the following key points:

- **Header**—Put your company name or your full name and perhaps a logo. Remember this is just a document for payment and will end up in the hands of accountants and not clients looking to hire you for your creativity.
- **Your Contact Details**—Include your name, company name, mailing address, phone number, and email address. If you have an EIN, it should also be included. Place this information on the top of your invoice and ensure it is easy to read.
- **Client's Contact Details**—Opposite your contact details, list the client's company name, address, phone, and email. Think of this as the "to" field since it defines who is expected to pay the invoice.
- **Job Title**—Include a short title that helps the reader know what job/production this invoice is related to. Remember the accountant paying your invoice probably sees many invoices daily. A simple description such as "Apple Launch Event September 2022" should suffice.
- **Invoice Number**—This will help you and your client identify the invoice to organize and communicate accordingly.
- **Date Prepared**—This date is important as it states when you created the invoice and can be used to assist when a client takes too long to pay.
- **Payment Terms**—A payment due date should be specified on the invoice. It may be listed in terms (e.g., 30 days) or have a specific date (e.g., October 18, 2022). The exact terms should have been agreed upon previously in your contract.

  - You may also detail any additional fees you charge if the payment is late. For instance, you could specify a late fee for non-payment by the due date. Some people apply a twenty-percent fee when the payment is beyond fifteen days late.

- **Breakdown of Services**—This area should be very detailed and describe the work and pricing charged to the client. You should list each day or week along with any specific overtime charges. If the client agreed to reimburse you for additional expenses, be sure to include those here too.

  - When detailing overtime charges, always indicate the date and the exact number of hours.
  - Provide receipts for reimbursement items along with your invoice.

- **Total Amount Due**—Do not forget to have a sum total amount at the bottom of your invoice and clearly indicate that this is the amount due. It should be easy for your client and their accountant to look at your invoice and see exactly how much they owe you.

Brad Schiller, Inc.

123 Fake Street, Austin TX 78758
EIN: 42-1010101010
512-555-1212
brad@bradschiller.com
www.bradschiller.com

Invoice

Bill To:

Homer Audio
742 Evergreen Terrace
Springfield, State 55555
614-555-1212

Invoice #: 0744

Date	Project	Sales Rep.	Terms
08/06/22	Simpsons Sales Meeting	Homer	Net 15

Description	Quantity	Rate	Total
Travel day – July 28, 2022	1	$XXX.xx	$XXX.xx
Audio Mixing – per day July 29 – August 2, 2022	5	$XXX.xx	$XXX.xx
Audio Mixing – overtime July 30, 2022	2	$XXX.xx	$XXX.xx
Audio Mixing – overtime July 31, 2022	2	$XXX.xx	$XXX.xx
Audio Mixing – overtime August 1, 2022	4	$XXX.xx	$XXX.xx
Audio Mixing – overtime August 2, 2022	5	$XXX.xx	$XXX.xx
Travel day – August 3, 2022	1	$XXX.xx	$XXX.xx
Per diem for daily meals and related expenses	7	$65.00	$455.00
Off-site airport parking in Austin, TX	7	$12.50	$87.50
Taxi to Hotel	1	$72.55	$72.55
Taxi to Airport	1	$75.24	$75.24

Balance Due: $1,000,000.00

Thank you for your business!

Figure 18.3 Sample Invoice

- **Thanks**—Some form of gratitude or appreciation for the work is always nice to place on your invoice. Just a simple line stating "Thank you for your business" can go a long way. You may also want to finalize the invoice with a statement such as: "Thank you, all work is complete." This makes it clear that you are no longer working on this specific project and that no other charges will be due to your client.

Invoices can be created in a number of different ways, such as on a spreadsheet, in an app, online, or in a business software platform such as Quick-Books. However you create the invoice, be sure to send it to your client as a PDF file. Alternately, you can print out the document and hand it directly to the client or even put it in the old-fashioned mailbox. Usually, however, a PDF as an email attachment is the best and the most preferred method. Alternately, some apps and programs will allow you to send a link to an online system where the client can view and pay the invoice. Always ask the client to confirm that they received your invoice, as even emails can be lost or skipped over.

Be sure to store copies of your invoices on your computer in an organized manner. You may need to refer back to them to confirm the amounts previously charged to the client or to send the invoice again when the payment does not arrive on time.

Other Business Documentation

While contracts, timesheets, and invoices are rather common, there are other documents that may be required by your clients. They may ask to see professional proof of insurance (e.g., for equipment or workman's comp). You might also be asked to sign (or you may ask a client to sign) a non-disclosure agreement. If you are working in a design capacity, you might have certain deliverables such as conceptual descriptions, music and sounds, drawings, cue sheets, and other show-specific items.

Make sure you understand the documentation that will be required of your position and organize your documents in a concise and consistent manner. Always take the time to carefully read the full text of anything you plan on signing. Ask questions about clauses you do not understand and never sign something that you do not agree with. If you have concerns about a contract or agreement, consult a lawyer or union representative for assistance. No job is worth signing something you do not agree with or completely understand.

19 Business Operations

Every position within the industry must adhere to some basic business practices. Most people work in some sort of capacity that could result in salary discrepancies, contract disagreements, cancellations, and other business operations. It is important to understand the basic principles and procedures to ensure smooth sailing through your career.

Getting Paid

At some point during your career, you will have a client or company that delays, forgets, or refuses to pay you for completed work. You have several options when it comes to collecting your money. These range from simply following up to taking them to court. It is always a good idea to start with the kindest method first.

First confirm that you sent your invoice, timesheet, and other documentation to the correct person and that their payment is late per the agreed payment terms. Looking back at your contract, invoice, and emails should easily provide you with these details. If everything looks good on your end, send your client a gentle email reminding them of the due payment and ask for a date they intend to pay.

If a customer continues to evade payment, reach out to the business's principal owner or manager. Moving up the chain of command can increase awareness and create action. Always be polite and use direct phone communication when possible. Clearly state the facts and ask for a quick resolution. After the conversation, follow up with an email to ensure that you have a paper trail.

If non-payment continues, you will need to prepare to take the client to court. Before hiring a lawyer and paying costly court fees, send a certified letter to the company letting them know your intentions to file a lawsuit based on the non-payment. Often you will find that this is enough for them to finally send your money. Otherwise you will need to decide if the court costs and the hassle of suing are worth it or if you're better off writing the situation off as a loss.

Although the goal is to have the client pay you the agreed amount, in some cases the best resolution may be to offer a settlement. Suggest a

DOI: 10.4324/9781003188957-22

payment plan or a reduced total payment. Some payment is always better than non-payment.

Pre-Payment Options

In certain situations, you may wish to negotiate a pre-payment to ensure that you receive the money you are promised. This can be very important with a new client or with a production that you think might be less secure than other employers.

For example, I was hired to work for an electronic music festival back before major corporations ran them. The producer was just some man that used to produce raves. I did not have much faith in his ability to pay me, or even be contacted, once the show was over. So I added further clauses to my contract requiring him to pay 50 percent of my fee 30 days before the event and the other 50 percent before the first act hit the stage. He promptly paid the first installment 30 days in advance, but his tune began to change on show day regarding the rest of my money. He kept promising to send a check the following week. I actually ended up in the production trailer with other crew members at the festival as bands were entering the stages. We all insisted on cash payment before we go back and perform our jobs. Finally, he gave us each the cash he owed us and we went back to our stations and provided a great show.

Pre-payment plans are certainly reasonable forms of doing business. In some cases, you may ask for early payment due to client concerns, or perhaps to simply help you financially before the event. Many designers ask for a portion of their fees in advance to enable them to earn an income while designing concepts and creating music and sounds. If you do plan to request early payments, always be sure to include this information in your initial contract or agreement.

Settling Contract Disputes

In some cases a client may not follow the guidelines outlined in your contract. Your process of resolving these disputes is very similar to that of non-payment. First, point out their error and ask for a correction of the situation. If they continue to not comply, then move up the chain of command. Always remain firm but kind. Do not resort to yelling or threatening statements.

Hopefully with reasonable discussions you can work out your dispute. In other cases, more drastic actions may be required. You may even need to resort to a lawsuit, but be sure to take all other reasonable actions first. With certain situations you could take your grievance to social media and speak out against the contract offender. Again, be sure to remain honest and kind while clearly presenting your grievance. Reserve this step as the last resort, as once your information is online, it cannot be taken back.

In some cases, you can leverage your yet-to-be provided work against their non-agreement to your terms. For example, a mix engineer worked for a major band for nearly twenty years. For all of those years, his contract stated that his international travel shall be booked in business class. The band's management had always complied. However, on a recent tour, the band decided they no longer wanted to pay any crew's business class ticket. It did not matter that the band was still on top of the charts and flying in private jets. When my friend found out that they would book him in coach for a twelve-hour flight, he refused to travel. The tour manager asked him to "take one for the team" and sit in coach. But he held his ground and continued to state that it was unfair and against his signed contract. As the date of travel loomed closer, the tour manager was finally able to gain clearance for him to fly business. The band eventually agreed that it was more important to have their usual mix engineer than to go against this particular contract point.

Cancellations

Events, shows, tours, or any gig can cancel suddenly for multiple reasons. It is always essential to have a clause in your contract regarding these situations. Once you book a production you will not be able to take any other work on those dates. If the production suddenly cancels, you will have missed out on other opportunities. For this reason, most production agreements have a statement similar to the following: "Cancellation of this booking within thirty days of the event will result in a charge of fifty percent of the rate for the booked period."

When a show cancels within your defined window, you should promptly send an invoice for the agreed amount and clearly state it as a cancellation fee. Most customers will honor your contract and follow through with payment. If you are working on a tour or a recurring production, you may need to alter the language of your contract accordingly to protect yourself regarding the period of time for lost work. For instance, if the lead singer in a band loses his voice and the tour ends two weeks early, you want to have a clause that ensures that you receive proper payment during that downtime.

When working for a union or as an employee for a company, be sure to follow the guidelines of your employer. They may require specific actions when a production cancels and you need to adhere to their standards. Always check the union or corporate policies regarding legalities and contracts.

20 Tax Form Basics

First, please remember that I am not a tax accountant or a taxation expert. Also keep in mind that your country or state may have different policies and laws than mine. However, it is important to understand the basics of taxation as they exist in the United States. You can then apply this knowledge to your local laws and procedures.

Paying taxes is a required part of life in most countries and should never be taken lightly. Be sure to consult an expert and always pay your taxes. Learn your local laws and procedures and follow them explicitly to avoid fines, fees, and possible jail time.

Self-Employed Taxation

If you are self-employed, the companies that contract you will have you fill out a W-9 form for the United States federal taxes. This is a basic form that simply provides your client with details such as your social security number or taxpayer ID and other personal information. They will use this data to report to the government the amount of money they paid you.

Then at the end of the year, the company will send you and the IRS a 1099 form that shows the IRS how much they paid you within that specific year. It is your responsibility to pay the taxes due on this amount. Remember too that just because your employer does not send you a 1099, this does not mean that you can skip paying taxes on the income you received. Always pay taxes on all received income, even if you do not receive a 1099! In the past, a form 1099-MISC has been used for all non-employees. Starting in 2020, form 1099-NEC (Non-Employee Compensation) has been reinstated.

In most cases, it is advised (and maybe required) to pay quarterly tax estimates so that you do not have to pay the full tax burden at the end of the year. If you are working through 1099s without any tax withholding, be sure that you are saving or pre-paying an estimate of what you will owe. You do not want to spend all your earned money and then have a large tax bill with no cash to pay for it!

DOI: 10.4324/9781003188957-23

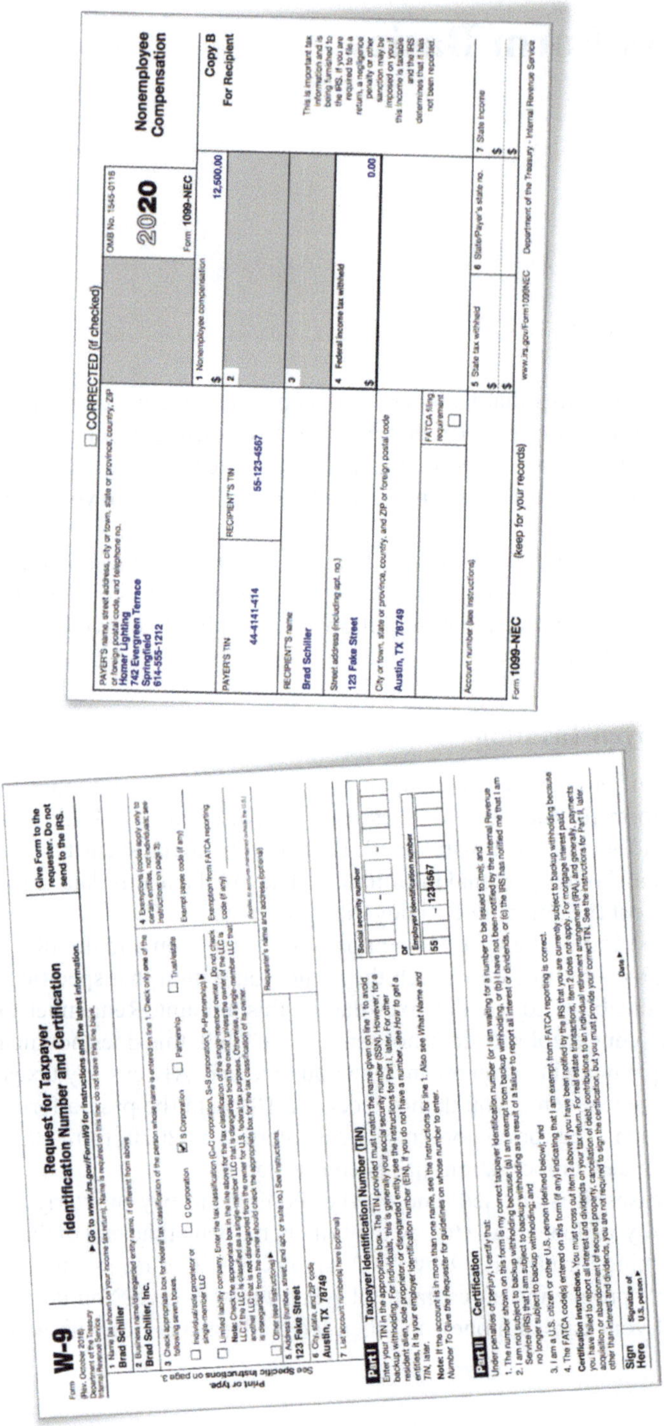

CORRECTED (if checked)

PAYER'S name, street address, city or town, state or province, country, ZIP or foreign postal code, and telephone no.
Homer Lighting
742 Evergreen Terrace
Springfield
614-555-1212

PAYER'S TIN 44-4141-414

RECIPIENT'S TIN 55-123-4567

RECIPIENT'S name Brad Schiller

Street address (including apt. no.) 123 Fake Street

City or town, state or province, country, and ZIP or foreign postal code Austin, TX 78749

Account number (see instructions)

FATCA filing requirement

1 Nonemployee compensation $ 12,500.00

2

3

4 Federal income tax withheld $ 0.00

5 State tax withheld $ $

6 State/Payer's state no.

7 State income $ $

OMB No. 1545-0116

2020

Form 1099-NEC

Nonemployee Compensation

Copy B For Recipient

This is important tax information and is being furnished to the IRS. If you are required to file a return, a negligence penalty or other sanction may be imposed on you if this income is taxable and the IRS determines that it has not been reported.

Form 1099-NEC (keep for your records) www.irs.gov/Form1099NEC Department of the Treasury - Internal Revenue Service

Form W-9 (Rev. October 2018) Department of the Treasury Internal Revenue Service

Request for Taxpayer Identification Number and Certification

▶ Go to www.irs.gov/FormW9 for instructions and the latest information.

Give Form to the requester. Do not send to the IRS.

1 Name (as shown on your income tax return). Name is required on this line; do not leave this line blank.
Brad Schiller

2 Business name/disregarded entity name, if different from above
Brad Schiller, Inc.

3 Check appropriate box for federal tax classification of the person whose name is entered on line 1. Check only one of the following seven boxes.
☐ Individual/sole proprietor or single-member LLC ☐ C Corporation ☒ S Corporation ☐ Partnership ☐ Trust/estate
☐ Limited liability company. Enter the tax classification (C=C corporation, S=S corporation, P=Partnership) ▶
☐ Other (see instructions) ▶

4 Exemptions (codes apply only to certain entities, not individuals; see instructions on page 3):
Exempt payee code (if any)
Exemption from FATCA reporting code (if any)

5 Address (number, street, and apt. or suite no.) See instructions.
123 Fake Street

Requester's name and address (optional)

6 City, state, and ZIP code
Austin, TX 78749

7 List account number(s) here (optional)

Print or type. See Specific Instructions on page 3.

Part I Taxpayer Identification Number (TIN)

Enter your TIN in the appropriate box. The TIN provided must match the name given on line 1 to avoid backup withholding. For individuals, this is generally your social security number (SSN). However, for a resident alien, sole proprietor, or disregarded entity, see the instructions for Part I, later. For other entities, it is your employer identification number (EIN). If you do not have a number, see *How to get a TIN*, later.

Note: If the account is in more than one name, see the instructions for line 1. Also see *What Name and Number To Give the Requester* for guidelines on whose number to enter.

Social security number

or

Employer identification number
55 - 1234567

Part II Certification

Under penalties of perjury, I certify that:

1. The number shown on this form is my correct taxpayer identification number (or I am waiting for a number to be issued to me); and
2. I am not subject to backup withholding because: (a) I am exempt from backup withholding, or (b) I have not been notified by the Internal Revenue Service (IRS) that I am subject to backup withholding as a result of a failure to report all interest or dividends, or (c) the IRS has notified me that I am no longer subject to backup withholding; and
3. I am a U.S. citizen or other U.S. person (defined below); and
4. The FATCA code(s) entered on this form (if any) indicating that I am exempt from FATCA reporting is correct.

Certification instructions. You must cross out item 2 above if you have been notified by the IRS that you are currently subject to backup withholding because you have failed to report all interest and dividends on your tax return. For real estate transactions, item 2 does not apply. For mortgage interest paid, acquisition or abandonment of secured property, cancellation of debt, contributions to an individual retirement arrangement (IRA), and generally, payments other than interest and dividends, you are not required to sign the certification, but you must provide your correct TIN. See the instructions for Part II, later.

Sign Here | Signature of U.S. person ▶ | Date ▶

Figure 20.1 IRS Forms W-9 and 1099-NEC

Company Employee Taxation

If you are working as an employee of a company, you will fill out a W-4 form when you are hired. This tells the company how much money you wish to have withheld from your paycheck for tax purposes. You will answer a series of questions about marital status, dependents, and additional possible deductions on this form.

Then at the end of the year, your employer will send you a W-2 form that indicates how much money was contributed to the federal, state, and Social Security and Medicare taxes. Depending on the amount withheld, you may owe additional tax funds or be entitled to a refund. Regularly review your W-4 selections and make changes as needed for optimum tax withholdings.

Pay Your Taxes

Regardless of your methods of withholdings, you will need to ensure that you properly pay your tax burden(s) by the required dates. This can be difficult when you are touring the world, but e-filing has certainly made it easier. Generally, you will need to fill out forms by entering data contained within all your W-2 and 1099 forms. Then follow the procedures to explain the amount of income you received and how much money was withheld (pre-paid). You can also subtract certain deduction amounts as determined by current tax laws. Any differences in the final taxation numbers will result in either you owing the government or them owing you.

This is a very simplified way to begin thinking about your taxation. The laws are ever changing and there are many things that you might be able to deduct to reduce your tax burden. For instance, if you are self-employed or own your own business and purchase specialized tools for your work, some of the amount paid may count towards reducing the total tax owed. Services such as websites, Internet, cell phone, accountant fees, and equipment are usually legitimate business deductions. The line starts to blur around home offices, mileage, meals, and other life versus business expenses. Always follow the law and consult tax professionals to ensure that you are adhering to the allowed deduction procedures.

Help Is Available

The majority of people working in the audio industry hire the services of a Certified Public Accountant (CPA). This professional will fill out your tax documents for you and possibly act as a business consultant regarding taxation and financial decision-making. A CPA should be up to date with all the federal, state, and local tax laws in order to ensure that you meet the legal requirements and look after your best financial interests. Most in the field find it best to hire a CPA familiar with the entertainment industry. While you likely will not find an "audio CPA," you can enquire with your local peers as to who they hire.

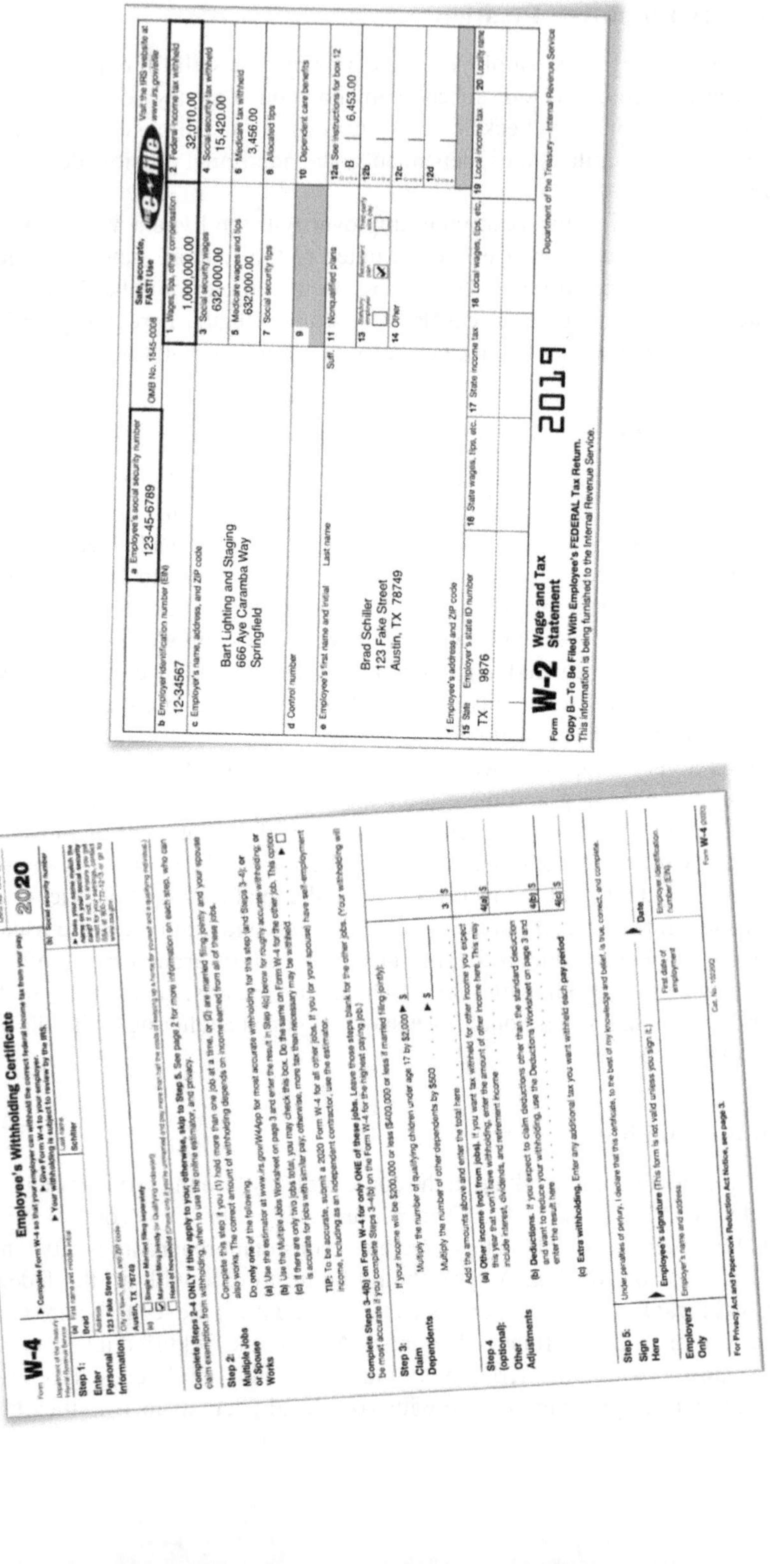

Figure 20.2 IRS Forms W-4 and W-2

Furthermore, some accountants specialize in working within the entertainment industry, helping musicians, actors, photographers, and others in similar professions. They are always a good choice, as they better understand the cycles and procedures of working in show business.

The IRS provides a wealth of information and forms for freelancers, employees, and business owners on their website (www.irs.gov). In fact, they have a dedicated section titled Small Business and Self Employed Tax Center, which can be found at www.irs.gov/businesses/small-businesses-self-employed. This site is a one-stop shop for all related documents, procedures, and information about the IRS and your business.

IRS Resources

The IRS has numerous documents available on various topics. The following are key IRS documents with which you should become familiar:

IRS Publication 334—Tax Guide for Small Business
IRS Publication 463—Travel, Entertainment, Gift, and Car Expenses
IRS Publication 505—Tax Withholding and Estimated Taxes
IRS Publication 535—Business Expenses
IRS Publication 583—Starting a Business and Keeping Records
IRS Publication 587—Business Use of Your Home
IRS Publication 936—Home Mortgage Interest Deduction
IRS Publication 946—How to Depreciate Property

Basic Documentation

It is essential that you track your income as well as outgoing expenses. Regardless of your business structure, this will be beneficial when taxes are due. In addition, it can help you see whether you are financially successful with your current employment choice.

When working as a freelancer, company owner, or union member, you will want to keep as much of your business finances separate from your personal funds as possible. The best way is to have a business banking account and a business credit card. With these tools you can keep money separate from your personal needs and expenses.

Next, you should track every dollar going in and out of these business accounts through a software program or app such as QuickBooks, Xero, FreshBooks, or FreeAgent. These tools allow you to document all your finances in one easy location. They even generate reports that can be easily sent to your CPA. Other tracking tools such as mileage logs, receipt scanners, and project tracking will come in handy for every freelancer. With all this documented information, it will be much easier to claim tax deductions and reimbursements where allowed and to prove the validity of your data in the event that you get audited.

When a company employs you (as opposed to you being a freelancer), you will receive your paycheck and not need anything further from a business financial reporting point of view. However, your company may ask for expense tracking and receipt scanning. If you are allowed to make purchases on behalf of a company, you must understand the Travel & Entertainment (T&E) policies of the company and learn their expense reporting tools. These systems are usually in place so that the company has the data required for their own taxation needs. Failure to follow their rules may result in you paying out of your own pocket for business expenses.

It is best to have a separate credit card specific to these business expenses; thus, many companies will issue you a card when you are hired. Be sure to get reimbursed for all allowed business expenses and never pay out of your pocket for a charge that the company should pay. Report your expenses honestly and always spend the company's money as you would if it were your own. Keep in mind that the expense reporting process is primarily for the purpose of aiding the company with deductions at tax time.

Taxation is a required result of earning money for everyone. Throughout your career, always pay your taxes and ensure that you understand the basics of taxation in your country. When needed, hire professionals to help with forms, requirements, and documentation. Remember that as your income grows, so too will your tax responsibilities.

21 Getting Gigs

One of the most common questions people ask when starting out in the live entertainment audio industry is: "How do I get more work?" Some of the best advice is to work hard, be nice, and talk with others. If people know who you are and what you are capable of, there is a chance that they might hire you. Of course, you need to have some skills and knowledge about the field as well.

Get Personal

Emailing, connecting through social media, texting, and submitting resumes are great, but this is a business of relationships and you must actually meet people face to face. Take the time to make phone calls, arrange meetings, and truly meet with professionals in the industry. If they have looked you in the eyes, there is a better chance that they will think of you for future work.

Networking is a key skill that will aid your career as much as audio skills and knowledge. You could be an extrovert and find networking simple, or an introvert struggling to connect with people. Many introverted personalities find it difficult to start conversations, but it is something you must do to gain career opportunities.

When you find yourself talking with other audio professionals, management, producers, or artists, be curious and listen. Ask questions and do not focus on selling yourself. Most importantly, follow up via email or phone call. As you openly discuss audio and life, you will find that opportunities may present themselves naturally.

For example, I was recently speaking at an event when a young person came up to me after my presentation. He asked some general questions about show business as well as specific questions related to my session. I then asked him what he was interested in and where he was currently working. He shared that he liked lighting, but video content was his area of specialization. He explained and showed me examples of his work on his computer. I was very impressed with his conversation and examples, so I mentioned that I had a need for some custom content on an upcoming show. After a few more interactions, I hired him to create and provide some imagery for my show. Had he never started a conversation with me, he would have missed out on this chance to expand and sell his work.

DOI: 10.4324/9781003188957-24

Not all conversations will lead to immediate employment, but the audio industry is rather small and you never know from where your next job will come. Furthermore, you can gain important knowledge from your connections with others that prove valuable for your future.

Get Out There

Creating a portfolio and/or website does not guarantee future work. You must get your name and skills in front of the people making hiring decisions. This will differ depending on the exact position you seek, but the basics of marketing yourself remain the same. First, realize that if you do not have much experience, or are relatively unknown, most people will not take a chance with you in a high position. You must earn their trust and respect regarding your knowledge and abilities and this only comes through proven hard work.

If you want to work more in concert touring, seek out the audio companies that are providing crews for tours. Get to know their staff, hiring managers, and crew chiefs. You may have to work in the shop for a bit to prove yourself, but it will be time well spent. Let them know your personal goals (e.g., eventually becoming a designer, engineer, system technician, or crew chief) and take gigs that support your future. You might have to meet with them on several occasions, take them out for coffee, or call repeatedly to remind them of your desire to work with them. Do not be too hasty though, as that could come across in a negative manner.

Perhaps you want to be a sound designer for television or theater. If so, going to audio shops is probably not the best place to start, unless you have zero experience as a designer. Instead, you need to associate yourself with producers, theater companies, television studios, and other designers. Again, you must determine who the people making the hiring decisions are and start networking with them. If these people know who you are and understand your abilities and experience, then they are more likely to think of you when a show needs a designer.

Whatever you desire for your future, there are jobs out there for you. You will need to research all the performing arts centers, houses of worship, production companies, audio shops, television studios, and other consumers of audio and then make contact with them. Opportunities abound in every city in the world, but the first step is simply getting potential employers to know who you are and what you can offer them.

Rejection Happens

Sometimes it can be frustrating to talk with people over and over when looking for work. Like in dating, there is often more rejection than acceptance. However, you will eventually find companies and events that wish to hire you. Until then, you must accept the rejections and continue pressing forward.

Before email was commonplace, I once mailed out a hundred copies of my resume looking for design work in the Dallas, Texas area. I blanketed the area's theaters, comedy clubs, production companies, theater groups, and venues. I thought that I would receive many offers for work. Instead, as the weeks went by, I became rather disappointed that no one was reaching out to me. What I failed to realize at the time was that there were three key principles working against me:

1. Jobs were not available right then, just because *I* was looking.
2. My work experience was very limited.
3. I did not follow up.

I just sat by the phone expecting work to come to me. Instead, I should have called each of the locations and asked them if they had received my resume. I could have spoken with someone and further explained who I was and what I could do for their organization. Even with my limited design experience, there were likely some companies that could make use of my skills. I never spoke with anyone directly to explain this and instead let only my resume speak for me.

Months later, the phone did ring, and I received one design job from that exercise. Had I followed up, I probably could have achieved several more. Remember that shows occurring soon likely already hired a production staff. You probably won't get put on a show in any position for some time after your initial contact with an organization. Depending on the timing of your employment query, there might not be any productions needing your talent at that particular moment. Jobs only materialize when organizations have a need for you and your talent.

Rejection can happen for many different reasons. Don't let it get you down; instead, use rejection as a lesson to learn how to improve for next time.

Get More Work

There is no easy answer to the question: "How do I get more work?" The best choice is to be a hard worker and make sure people know who you are and your capabilities. Then network with every possible person to become known in the circles within which you desire recognition. Don't be in a rush to the top and be sure to always follow up with others. With these basic skills, the jobs will come your way.

22 Planning for Your Future

At the beginning of their careers, most people are not ready to think about retirement. It is usually 30 to 40 years away and it's exciting to put the income from a new job to current needs. However, it is essential that you plan ahead in some fashion for your future. Inevitably, there will be a time in your life when you wish to work less or not at all. Proper financial planning now helps you live comfortably without income concerns when you slow down or retire completely from working.

Generally speaking, most experts and financial companies agree that you should save approximately 10 to 15 percent of your annual income for retirement. In theory, it is fairly easy to stockpile money for later on. For instance, if you simply invest $200 a month for 48 years, you will amass over a million dollars. This calculation assumes a seven percent average annualized growth in a tax-deferred account. Forty-eight years is a long time, of course, but if you are currently in your 20s you will likely be employable for the next 40 or more years. Additionally, if you increase the amount saved per month, then the time duration to save a million dollars will decrease dramatically.

While saving $200 a month sounds like a great plan, it usually is not realistic for people starting out. If you have significant debt such as student loans, rent or mortgage, car payments, credit card bills, and insurance, then you might not have an extra $200 a month to save. Furthermore, if the savings interest is less than your debt interest, you will not be helping your financial situation by stashing extra cash away. In many instances, it is best to pay off major debt first before investing in your retirement. Remember though that it never hurts to start a habit of saving as early as possible.

The Best Path to Wealth

The very best way to ensure financial security, both in the present and the future, is to earn more and spend less. As you earn money throughout your life, you may desire bright, shiny, and new items. In most of the world, much of society is focused on material objects and ownership. Do your best to avoid reckless spending and avoid debt as much as possible. Never purchase things you cannot afford and live within your means.

DOI: 10.4324/9781003188957-25

As your career grows, you should strive to increase your earning potential and actual income. If you have been an audio technician for five years, making the same amount of money per week, then it is time to increase your rate. Look for opportunities throughout your career to boost your income. Usually you will have to ask for an increase, as employers are always looking for the best deal they can get when paying employees. Take responsibility for your own income and ask for a raise or promotion every few years.

I once worked for an employer who did not offer raises and preferred to keep employees at their same salary levels year after year. I worked for this employer for quite some time, increasing my knowledge and value, yet they did not financially compensate me for becoming a better employee than I was when they hired me. Furthermore, due to economic inflation, the salary I was paid was actually worth less than when I started. A smart friend pointed out that I was actually being paid less and less each year (in terms of the dollar value). I finally went to the HR department and asked if they thought I was worth less to them now than when I started. I also asked for a raise and negotiated for "cost of living" increases moving forward.

Take the time to ensure that you are being paid a fair wage for your current role, and as you grow in your career, ask for increased compensation for your services. Over your lifetime, your income should increase steadily and provide you with more opportunities to save for your retirement. Try not to spend all your newfound income and always be diligent in your savings plans.

Actual Retirement Savings

When it comes to saving for retirement, there are many choices available to those of us living in the United States. Many other countries have similar options available and it is important to research the possibilities within your country. Furthermore, the assistance of a financial advisor will certainly prove beneficial towards any retirement plan.

As an Employee

If you are an employee of a company or a union member, then you probably have some sort of retirement plan available to you. You can elect to have a percentage of your income deducted from your paycheck and placed in an investment of your choice. Many 401(k) plans allow you to select the type of investment for your retirement dollars. Take the time to study the options and select a strong choice. Furthermore, the money going towards retirement savings is usually tax-deferred, meaning you won't pay taxes on this income until you withdraw it at retirement age. The amount you contribute per check is up to you, but generally it is a good option to save ten percent or more of your income.

If your employer offers a matching program, always collect as much of this as possible. It is free money that is yours if you simply elect to save a specified percentage of your income for your future. For instance, assume your employer will match 100 percent of the first three percent you invest. If you are earning a salary of $60,000, this means that your employer will add $1,800 into your retirement savings annually (assuming you also save at least three percent). Remember that there are few regulations regarding matching and employers can create these programs as they wish. Some will match within certain constrains or caps, and most will also require investing via selected methods. Often you must be employed for a certain period of time before the matched dollars actually become yours. If you leave the company before this timeframe, you will not keep the matched investment dollars. Always take time to fully understand the retirement program offered by your employer and ask questions to the HR team.

If you change employers, transfer your 401(k) dollars from the previous employer to your new employer's plan. There are many rules and regulations regarding the transfer of retirement funds, so be sure to educate yourself and get help as needed. Improper transitioning of plans could result in fines and/or taxes due.

As an Individual

If you are working as a freelancer or an owner of a small company, you still need to save for retirement. Obviously, an employer-based 401(k) will not be available, but you do have several good options for creating your own retirement savings plan.

Individual Retirement Account (IRA)

An individual retirement account (IRA) is available to anyone who has an income, whether self-employed or working for a company. The government sets and adjusts contribution limits. Currently, if you are under 50 years old, you can contribute up to $6,000 a year to a Roth or traditional IRA. If you are over 50 years old, then you can contribute $1,000 more annually. There are two types of IRAs: Roth and traditional. The key difference between the two is that a Roth is taxed annually, while a traditional IRA is taxed when withdrawn. Contributions to an IRA are tax-deductible and will help to reduce your tax burden at the end of the year.

When you are younger and earning less, it is a good idea to invest in a Roth IRA as you will be taxed at your current tax bracket. In the future, you will hopefully have a higher income and thus be in a higher tax bracket. Then a traditional IRA may be a better choice as the tax is deferred until retirement.

Simplified Employee Pension (SEP)

Anyone working as a freelance employee is eligible for a Simplified Employee Pension (SEP) IRA. This investment option allows you to contribute up to 25 percent of your freelance income or $58,000 a year—whichever is less. You can fund it up until April 15 of the following tax year. This enables you to determine how much you wish to save after you have completed calculating your income for the previous year. Also, with a SEP IRA, you are not required to make a contribution every year, which can be helpful when life happens and you wish to spend your dollars immediately instead of saving for retirement.

Solo 401(k)

As the name suggests, this is a 401(k) plan for an individual. There are limits to how much you can contribute individually, but you can also make contributions as an employer. So you can increase the individual annual limit of $19,500 to $57,000, depending on the IRS rules and your income. Solo 401(k) plans are set up as either pre- or post-tax and often include additional fees for investment companies as well as complex IRS paperwork requirements.

Even More Options

Beyond traditional funds and government programs, there are additional ways to save for your future. Real estate, businesses, stock markets, certificates of deposit, and other investments can yield great returns. I know of many in our business who own multiple rental properties. Not only are they earning monthly income from the renters, they also hope to receive a hefty return on the sale of the property in the future. Others have invested money in various business ventures such as audio companies, rehearsal studios, restaurants, car washes, and equipment.

Diversifying your savings with various types of investments is usually a good idea as you will average out better if something goes wrong with one of your choices. Don't go into debt to form an investment; rather, take the time to save money first and then invest wisely.

Government Programs

The Social Security program in the United States allows workers to save money for retirement. At the age of 62, you become eligible for retirement benefits and can begin receiving a monthly stipend from the government. Other countries have similar programs in which residents contribute a portion of their income to the government in exchange for benefits after a certain age. While most programs are required by law, some are also optional. Always check your local laws and procedures to ensure that you are in compliance and eligible to receive benefits when available.

A Helpful Resource

At least one company provides financial education and resources to employees in the arts and entertainment industry. The Institute for Financial Wellness for the Arts (www.theifwa.com) has training courses, calculators, information portals, and links to financial coaches. Their services are available to both individuals and companies. They can assist you in your retirement planning as well as daily financial decisions. Several major financial companies such as Fidelity Investments, Charles Schwab, and Vanguard provide retirement planning and assistance without a focus on the entertainment industry.

Have a Great Life

Throughout your career, your retirement needs, choices, and abilities are likely to change. It is vital that you continually plan for your future with retirement savings. You should evaluate your plan periodically and decide if it still suits your perceived needs. As you move between employers or to and from freelance positions, always keep saving. Talk with a financial advisor about your best investment options and limit your spending to only within your means.

With a focus on your future, you can live comfortably now while also saving for later in life. Unfortunately, some in the entertainment audio industry have lived very successful professional lives without planning well for retirement. This has left them struggling later in life. On the other hand, many colleagues have saved and will be very comfortable after retiring. Choose now for a better life by saving for your later years. With a solid financial backing, it will be easier to decide when to slow down or retire completely from working.

Part 4
The Creativity

Entertainment audio provides for many fun and exciting careers. Whether you are involved as a technician, designer, mixer, system engineer, or any other position related to entertainment audio, you will find that your creativity is beneficial to your craft.

Many people are born with natural creativity, while others have to work at it. The entertainment audio industry is unique as it allows you to utilize both your technical and creative talents. The very core of a sound designer's job is highly creative, but they must also possess technical skills to develop and communicate the acoustics, concepts, and sounds desired for the show. A sound

DOI: 10.4324/9781003188957-26

mixer provides creative input as they adapt the EQ and balance of the sound according to the production's needs. However, the mixer's job of manipulating the desk is also very technical. An audio technician utilizes technical skills to repair and operate equipment while their creativity adds expertise to speaker layouts and uses. By combining your technical and creative skills, you can achieve exciting and fun audio for productions.

If you don't feed and exercise your body regularly you will become overweight and tire easily. In the same manner, if you do not frequently feed and exercise the creative portions of your brain, you will become a non-creative, boring artist who fails easily. Audio is an art and art requires innovation and expressive freedom that only come from new thinking and ideas.

Each position in entertainment audio provides many avenues for creativity. However, it is also easy to rest on your laurels and continually reuse the same concepts, procedures, and ideas. Work to free your mind from the traps of stagnant thought and become more creative as you approach each production with new understanding and insight.

23 Abundant Creativity

Each role within the live entertainment audio field has creativity to contribute. Whether you are a sound designer or part of the shop staff, there are opportunities to be thoughtful in new and distinctive ways that can greatly enhance a production or routine.

Artsy Sound Design

A sound designer is hired by the artist, director, or producer to craft the creative approach to a production's sound. Of course, the audio also needs to be functional, allowing the audience to hear the event occurring on stage. But beyond amplification, sound can provide an additional creative dialogue for a show. Ambient sounds, creative effects, realistic environments, and direction of sounds can convey emotion, attitude, and time, as well as other concepts. The sound designer must determine how to best use entertainment audio to enhance the production without compromising budget, time, and other valuable resources. To be successful, a sound designer must be creative and innovative throughout their career. Only by approaching the same problem with new solutions can a designer truly provide a production with a creative use of entertainment audio.

Creative Mixing

Mix engineers are frequently the driving force of entertainment audio. They command the sound on a daily basis and must be familiar with not only the abilities of the console and speakers, but also the subtleties of sound of the music or spoken word. As a mixer it is very important to be creative while maintaining the technical accuracy required when mixing an act on stage. Understanding the needs of the show, the band's preference for sound, and/or the purpose of the mix are key elements upon which the mixer must apply their creativity. A purely technical mixer is simply adjusting the volume of the various inputs with no regard to the creative needs of the show.

DOI: 10.4324/9781003188957-27

However, the mixer's creativity must not override that of the band or the performance. Ultimately the sound should be composed in a way that is best for the production or the act. This proves a difficult task for many and this is why you will frequently see a mixer working with the same artist(s) across multiple productions.

Motivated Technicians

The live entertainment audio technician plays a highly important role in a production by installing and maintaining equipment. Without proper equipment maintenance, the production will suffer, as the audience will not properly hear the creative efforts of the designer and/or mixer. The tireless work of technicians ensures that productions are repeatable for audiences, day in and day out. One might question how a technician can provide creative concepts or "think outside the box" to help a production; however, these actions are essential to the success of a production. Technicians regularly contribute artistic ideas for speaker or microphone placement, rigging, and system usage. A designer or engineer may not be aware of certain limitations nor have the abilities that only a technician can achieve through specialized knowledge of the equipment and the specific sound rig. Technicians should speak up and suggest new methods for entertainment audio practices to further enhance the creativity of a production.

Let's All Work Together

An entertainment production requires many skilled craftspeople to fully achieve its audio needs. Only by working together to provide the best possible show can all the expertise converge. Open sharing of ideas helps to inspire others to produce unique solutions and innovative concepts. Everyone involved with a production needs to feel free to contribute ideas and thoughts. Even if a specific suggested concept is not utilized, it might inspire further notions that ultimately benefit the show.

Speak up and share your ideas with others and see what transpires. Creativity sparked from one person will often ignite a passion in others. Whatever your position, verbalize freely your creative ideas and see where they take you. Of course, you don't want to criticize the efforts of others as you make suggestions, so use caution with your phrasing. Once ideas are on the table, be open to modifications or suggestions and always keep the best interest of the show in mind. When we all work together, we can help create the best show ever for every production.

24 Creative Inspiration

Creative inspiration can come from various sources, not just from our minds or coworkers. The creative aspect of working with entertainment audio involves continuous study of countless different resources and processes.

Show Biz

The most obvious place to look for the creative process of entertainment sound is in other productions. Many find it very important to listen to different types of productions and to try to remove their "production" hat. That is, they try to watch and listen from a *normal* audience member's point of view. This is not easy, of course, because we naturally want to dissect the technical aspects of the show. Over time, though, most get better at simply watching and listening to the entirety of a show without continuously thinking about its technical bits.

Streaming media, DVDs, and televised events are good resources for reviewing shows, but remember it's best to hear a production in the medium in which it was originally created. The audio in a concert tour video stream or recording always sounds different from its live, in-person concert version. I suggest attending shows of all different sizes and types. Go see the big shows in Las Vegas and on Broadway, as well as popular concert tours, but also check out a local high school production or community theater show. You will see that limited resources often enhance the overall creative output. Also remember that ticket fees are often tax deductible as research expenses.

Natural Stimulation

One of the best inspirations for sound comes from nature itself. Even though you may not try to replicate the exact sound of a waterfall, bird chirps, or thunder, most find the randomness and beauty in nature more inspirational for their own creativity than any other source. Make it a practice to listen deeply to the world around you and determine how sounds are created and manipulated naturally without any human or technological interactions.

DOI: 10.4324/9781003188957-28

Figure 24.1 Nature Provides Many Sources of Information

Echoes, reflections, tone, balance, volume, pitch, duration, frequency, dynamics, and so forth are all readily available to be studied just outside your window or venue doors. Simply spend a little bit of time focusing on the environment around you and noticing the intricate details. You will be amazed at the increase of your creative potential during your next audio production.

Whether working on a theatrical production, corporate event, or live music tour, you can find a way to implement the features of natural sound into your event. Perhaps an echo in a canyon will inspire you to add a bit of reverb to the mix or the sing-song of a bird will entice you to emulate its rhythm within a sound effect for a show.

Creative Writing

I recently did a search for books on amazon.com for the word "creativity" and found over 80,000 titles. While they may not all be winners, there are many wonderful books about triggering and exercising your creative talents. In fact, you are reading a great source of creative inspiration at this very moment!

Many years ago, a designer told me that he reads *A Whack on the Side of the Head* by Roger von Oech once a year. I have this wonderful book, and while I do not read it that often, I do try to read it once every few years. The book's simple exercises modestly point out that we should not get trapped into the same ideas and methods, and instead should "whack" ourselves upside our heads to help develop new ideas and concepts.

Creativity Is Just One Talent

As stated before, you can enjoy entertainment audio for both its technical and creative aspects. Creativity comes in many forms and can be applied regardless of your current position within the audio field. Throughout your life, you have likely heard the term "think outside the box." Only by actively improving your creativity and applying different lessons to your work can you truly become successful at providing innovative audio and solutions.

Part 5

The Lifestyle

Working in the entertainment audio industry is often a career of passion more than a career of financial wealth. While not producing as many millionaires as other professions, the field does allow workers to have wonderful careers, nice salaries, and great work-life balance. Although there might be difficulties related to relationships and health, working in entertainment audio is generally a wonderful choice.

DOI: 10.4324/9781003188957-29

Part 5

The Lifestyle

25　Why We Do It

The majority of the people working in the live entertainment audio industry are doing so to satisfy an internal passion and not to earn millions of dollars. If your desire is earning heaps of money, this field is probably not for you. However, if you want to really enjoy your work and make a good living, then you have come to the right place.

Sound people usually start off working as a member of the technical crew in high school or come from being a musician in a band. The majority find a calling in audio and many feel that "sound is either in your blood or it is not." The joy of being part of a production, amplifying actors, augmenting a band's performance, and bringing emotion through sound are only the start of what drives most. Perks include getting to work with amazing people and state-of-the-art technology, traveling the world, and helping to entertain others. Not one sound person has ever created an entire show alone, and being a part of a larger team is very satisfying on many levels.

Some want to work as artists and find that the art of audio matches their creative desires. Others just enjoy the technical aspects or the thrill of bringing entertainment to audiences. A few even satisfy their dreams of being performers, but achieve this in a technical way rather than by focusing on a personal performance skill.

Backstage magic happens endlessly around the world and many find opportunities to be a part of the crew very enticing. Audio enhances productions, allows followers to hear their idols, and entertains directly with unique soundscapes and effects. Whether working as a manufacturer's salesperson, sound designer, or audio technician, each person plays a pivotal role in the success of a production.

A very good financial living situation is possible to achieve while working within the audio field, regardless of your position. You do not need to become a top sound designer or mixer to live a very comfortable life. Many find the audio field a profitable environment from which to live. While only a few at the top can make millions of dollars, most within the industry are able to afford all they want as well as save for their retirement. Having a quality life while doing something you love is a true marker of success.

DOI: 10.4324/9781003188957-30

Work Hard, Play Hard

Putting a show together involves lots of hard work, both physical and mental. People in all positions and levels of the industry work very hard to bring audio to productions. If you have the passion, you will find joy in the hard work and not find the mundane tasks a burden. Many people find that they are in the "zone" (or flow state) when doing their work with sound. I believe this to be a key factor in a successful career.

Often, when working on a drawing, operating a console, or simply loading in or out, I lose track of time and stay present with the duties in front of me. This mindset is commonly referred to as a flow state and is often achieved by people who are totally engrossed in the moment. Experts agree that this condition is good for our minds and wellbeing. Therefore, the more we can achieve this frame of mind, the happier we will become. Whenever you enter the flow state during your primary work, you know that you are following an internal passion.

You must enjoy the majority of the work that you are doing; otherwise, it just becomes labor. Sure, you might not want to have to roll up that FOH snake after an outdoor festival, but you do it anyways. The inner passion to be a part of the production allows you to gather the strength and desire to get your hands dirty and roll the cable. Not every single task within audio will be fun or even easy for you.

Besides, at the end of the load-out, you know that your friends are waiting for you. By friends, I mean the rest of the sound crew. Most lighting people stick together and actually enjoy spending time with each other. Like-minded people tend to enjoy each other's company. Those working together on the audio of a show create many friendships and working bonds. The industry is based upon relationships and most of these are formed and solidified while working actual events.

Limited Stress

Generally, the industry provides mostly low-stress positions of employment. Sure, there can be major stress to get a show's sound finished on time and on budget, but we are not trying to save human lives. We are working within the entertainment industry. It is very important to remember that most productions have a limited run and thus will be over at some point. New tours, productions, and products will be created and old ones will be forgotten.

However, stress does materialize regularly and is usually related to relationships with others, production politics, or limited time. Learn to recognize stressful situations and try to minimize the increase in your personal pressure-level. Take a moment to think about the situation and then respond, instead of just reacting. When an irritant presents itself, don't react to it; always wait a few seconds or minutes for a better response to come to mind. This will lead to smoother conversations, improved workflows, and lowered stress.

When in difficult circumstances, I will often sing to myself the first few lines of the song, "There's No Business Like Show Business." This calms me down and reminds me of the joy and the purpose surrounding the work in which I am currently involved. In fact, take a moment to look up the full lyrics of the song to gain further show business inspiration.

Because the stress in the field is limited and production-based, most enjoy their careers with relatively minimal work-related trauma. If you can learn to see the stressful situations as fleeting, you can start anew with the next production. There are enough stressors in our personal lives that we do not need more from our careers. Luckily, the live entertainment audio field provides generally low-stress jobs for most. By releasing the burdens of previous shows and managing your time, you can lead a low-tension life.

Unbelievable Situations

We are very fortunate to work in an industry that is built upon helping people find joy and happiness. The entertainment field affords its professionals many opportunities that are not always available to those working in "normal" jobs. Often, we experience some of the high life as we work alongside celebrities. Furthermore, the very nature of the entertainment lifestyle brings us into locations and situations that may seem unbelievable to outsiders.

Tour buses, private planes, five-star hotels, exclusive venues, fine food and drink, private performances, and unique locales can be yours when working in the entertainment audio industry. I have exited a limousine and walked the red carpet at the Oscars, flown on a band member's private Lear jet, slept in presidential suites, been escorted by police motorcades, visited top-secret sites, and experienced many more incredible occasions. These exciting moments may not come early in your career, but they are certainly possible regardless of your position or role. These occurrences abound and will happen when least expected. Furthermore, sound happens all over the world, thus paid-for travel to any place on the planet is also possible (and very common).

When traveling for work, try to take some time to actually see the location you have arrived to. Sure, there will be instances when you only see the venue, but make the most of your time and get out and see the city whenever possible. You can see quite a few sites in a town over the course of just a few hours. When you can, schedule an extra day to explore the locale to which you have been sent. If you are on a production and have a day off, get out of your hotel room and see something in the local area. Why would you ever skip going to see the Eiffel Tower when someone else has paid for your trip to Paris?

Audio Is Life

A career within the live entertainment audio field will provide you with a good lifestyle as well as amazing opportunities that you never thought possible. The rush that you feel when working on a large production and thinking back to your high school beginnings is unimaginable. If audio is in your blood, you will automatically strive to do all you can to grow your career in a way that suits you best.

26 Work-Life Balance

We all want to live balanced and healthy lives. However, the desire to grow and participate in our careers can often take a toll on other aspects of our lives. Our lives consist of more than just our work; our lives also comprise important facets such as family, community, spirituality, and social life. Learning to balance your work with all of these aspects of your life and knowing how to adapt to life's situations is vital.

Workers in any industry might fall into situations in which their work detracts from their personal lives, resulting in many depressing and devastating situations. Unfortunately, the audio industry is not immune from this type of pitfall; in fact, the work often requires more time away from your family and personal life than common professions. It is very important to find a way to balance your work and your personal life, even if you feel that your work *is* your life.

Working Hours

First, realize that the live entertainment industry generally works during the hours that others are looking for enjoyment. Quite often, this means working nights, weekends, holidays, and other untraditional hours. This will have an impact on your social, spiritual, and familial lifestyles. Interacting with your friends and loved ones will become difficult, as they plan events and gatherings that conflict with your work schedule.

Audio departments are sometimes asked to work when others from the production are not present. This is because they are using loud test noises, making it difficult for carpenters, performers, band members, and other crew to safely accomplish what they need to do. Therefore, you are sometimes relegated to working the overnight hours after the others have completed their time on stage. Then you must readjust to the hours of the production's rehearsals and shows. Adjusting to these shifts in your working hours can take heavy tolls on your mind and body, as well as be confusing to your family and friends.

Remember too that the show is going to happen with or without you. You are just one part of the entire production. Your needs to schedule around personal events just won't matter to the overall production. Instead, you will generally

DOI: 10.4324/9781003188957-31

need to schedule your life activities around the needs of the show. Asking for a night off to take your spouse to the movies does not happen. It is more than likely that you will need to arrange your personal events according to the free times in the production's schedule.

Growing Your Career

When working as an independent freelancer in the audio industry, you will want to say yes to as many opportunities as possible. Often this will mean missing birthdays, holidays, celebrations, and social activities. If you are offered the chance to mix the sound for a band you have been courting, you probably will not turn it down just because of a single night's personal conflict. Instead, you will take the gig and miss that particular occasion in your life.

As your career grows, it should become easier to selectively say no to conflicting events, but this is not usually the case. Most build a work ethic of saying yes to gigs above all else. This routine then creates difficulty when making decisions between life events and work opportunities. Most people will default to work choices and let their social and family lives suffer. However, there are many times when it will be more important to select your personal life over career moments. You need to weigh each situation carefully and make informed choices. There will always be more gigs, but many personal events won't ever happen again.

I Love My Work

As stated earlier in this book, most enter the field with a strong passion for working with audio. This leads to a drive to work more shows because the work does not feel like labor. It just feels like a part of life and generally is lots of fun. This outlook leads to a wonderful lifestyle as you don't dread working. However, it can also lead to confusion and depression when it comes to choosing your passion over your personal life.

There are certainly times in your life when you should focus a large part of your energy on your work. However, as relationships foster and families grow, the responsibilities outside of work increase exponentially. The decision-making process will become more difficult. Leading a life consisting only of audio work is not a complete life, even if you absolutely love what you are doing.

Finding the Balance

Having a wonderful, well-rounded life means living with a balance between your professional and personal lives. Believe me, this requires effort and discipline, along with an understanding from the people in your life. You need openness and honesty in conversations with your clients, coworkers, family, and friends to maintain the balance.

Whether traveling or just working wacky hours, there will be many situations in which you need to choose between employment and personal moments. Missing life's activities such as birthdays, school events, social commitments, and personal time so that you can earn money to support yourself and your family is a very commendable mission. Nevertheless, you must also weigh the emotional toll on yourself and others when choosing work over other parts of your life.

Once while working a show, I overheard the sound designer make a phone call while sitting near me at front of house. He called his young son and wished him a happy birthday. I heard him apologize for not being there and ask if his son had received the gift he had shipped him. It became clear that his son was more concerned with his father's absence than his gift. When the call was over, the designer was noticeably shaken and needed time to recover from the call.

In that moment, I thought of my own son who was only a year old at the time. I decided right then and there that I did not ever want to miss his birthday. I resolved that I would do all I could to ensure that I was never in the same situation as this designer. I made a vow to myself that I would not work on his birthday until my son was old enough to say, "Dad, I am doing my own thing for my birthday." For the next fourteen years or so, I was always present for his birthday celebrations and not away working. I was lucky that this meant turning down only one gig, as for the rest of the time I found ways to ensure I was home without turning anything down. I am forever grateful to that designer for teaching me this lesson and motivating me to stick to my convictions.

Personal discipline is key to a good work-life balance. You must check in with yourself from time to time to ensure that you are happy with the choices you are making. If you feel that you are missing too many social events, make a change and turn down a show or ask for some time off. Although the entertainment business works many long strange hours, it is also flexible when it comes to life. After turning down a show for personal reasons, you will find that more opportunities come your way. Don't let the drive for more money or experience take away from life's precious moments with those you love, including yourself.

Relationships

Personal relationships, marriages, and family connections are the most common parts of our lives that must be balanced with our work. Open and frequent communication is very important, and luckily, modern technology makes this a bit easier to achieve than in the past. To find a balance, communicate with your loved ones often when you are away working on projects. Take the time when you are walking to catering to call your child or spouse. Instead of going to the bar after the show, choose to FaceTime or Zoom with your parents. It is too easy to simply focus on our jobs and put our family aside until we return. But it is important to keep our families in the loop with our lives.

Maintaining a romantic relationship is very difficult under many circumstances, and the industry requirements certainly do not make it any easier. We are often traveling, working in environments of excitement, and meeting many people. Those who remain at home are often unclear about what exactly goes on backstage and form their own thoughts. Some of the stereotypical temptations are real and others are imaginary, but that does not make them any less of a concern. Couples must take the time to be transparent about their lives with each other and to correspond honestly. When a partner lets you know that they need you home, make time to be home. A good companion will also understand your desire to work more and provide you the freedom to work as much as you wish.

Unfortunately, some romantic relationships in the industry do not survive very long due to the rigors of the road. Usually there are numerous other factors that contribute as well, but a big ripple comes from time spent apart. Do all you can to balance your absences with phone calls, texts, emails, and other methods of comfort. Never take your loved ones for granted and always let them know you are thinking of them. Small tokens of sentiment go a long way during times of absence.

Life Happens

Throughout your career, both good and bad circumstances will occur in your life. Some will affect your career directly and others will transpire with little to no impact. When life's situations do take place, be clear and honest with those you work with and ask for the help you need. Everyone experiences deaths, family events, celebrations, personal troubles, and relationship concerns. If you are working a show and get unexpected news, do not be afraid to let the production know what happened and indicate that you need to leave. The industry is very understanding and helpful during these occurrences.

For instance, I was on a tour and my father was selected to receive a very important honor. I really wanted to be at the ceremony to share this momentous occasion with him. I spoke with the production manager of the tour and discovered that the event date conflicted with only one show during the tour. Furthermore, the tour stop was a festival day. I was allowed to fly to my dad's event after the previous show and return to the tour after his award ceremony. I simply had to teach someone on the crew to take my place at the console during my absence. I already had a backup person, so this was easy to achieve. While I was exhausted from the extra travel, my family was thrilled that I was able to make it to the event and the tour easily continued during my absence.

I have heard of similar stories for sudden unfortunate events too. Trust me that no show is worth staying at if it means missing important life moments. Death and accidents happen throughout everyone's life, and in most cases you need to choose to be with your loved ones during these times. You cannot get missed time back, and the show will always go on, with or without you. With open dialogue among all concerned, you should be able to balance your career responsibilities along with life's events.

Take Care of You

The best way to find a healthy work-life balance is to check in with yourself as much as possible. If you feel that you are missing out on personal events, self-practices, relationships, social engagement, or other parts of life, then dial down your employment responsibilities. Have open discussions with loved ones and coworkers to help you maintain a balance. Even reject work from time to time to focus on other aspects of your life. There will always be more jobs, but many of life's occurrences won't ever happen again.

If you feel that life is getting out of control and that balance cannot be achieved, please look for some assistance. There are many ways to grow personally including books, seminars, therapists, coaches, and charities. Help is available and you simply need to decide to start improving your life. A good career does not only involve working with sound; it is also important to have a balanced life with meaningful relationships, social activities, and strong connections. Be sure to balance yourself often to ensure a well-rounded life and career.

27 Allure of Celebrity

Working in the entertainment industry allows many interactions with celebrities of all types. I have had the pleasure of meeting rock stars, actors, models, CEOs, politicians, and even an astronaut who walked on the moon. Celebrities come in all types, and working with them presents some unique situations and challenges. But what, or who exactly, is a celebrity? An online search defines the word celebrity as "a famous or well-known person." Anyone from a musician to a business icon can be a celebrity. When working within the entertainment audio industry, you will undoubtedly work near and alongside several celebrities. Regardless of your role, it is important to understand the characteristics and expectations of working with famous people.

Working With the Famous

Non-industry friends often ask me if I get to meet famous people and what it is like to work with them. I think they expect me to give some sort of tabloid gossip to further explain the behaviors of a rock star; however, this is not what I share. I usually explain that working with a well-known person is no different than working with a not-so-well-known person. We are each an important person on this planet. Just because millions adore one person, and another is known by only a few, it doesn't mean that there should a difference in how you work with them.

When a "star" comes to a production meeting or visits FOH, I treat them the same way as I would any other professionals I am working with. I respect that since they are the "artist" they can make certain decisions, but this is no different than an office staff having a meeting with their CEO. Quite often the celebrity is a very creative individual who has some interesting ideas to share. Their years of entertainment experience has created a unique perspective that often leads to specific desires for audio equipment or sound design. It is important to treat these celebrities as individuals and not show them preferential treatment or considerations. In other words, don't kiss their ass just because they are famous. They get plenty of that on a daily basis, and most just want to cooperate to get the show to a perfect state.

DOI: 10.4324/9781003188957-32

Respect Their Privacy

I happen to have the cell phone numbers and email addresses of some of the celebrities who I worked with. This personal information was provided in trust based on our working relationship. I respect their privacy and do not contact them to simply chat about the weather or impress my friends. I am merely a working professional who has been a part of the same productions as them. We are not buddies, so I don't feel I need to call any of them just because I know their numbers. When we are working on a project and need to speak about the show, I will make contact; otherwise, I refrain from arbitrary communications.

Furthermore, your website and social media should not contain photos of you hanging out with famous people. You are not in their company as a fan. You just happened to have worked on productions that included a famous person or two. I have seen some production staff post photos of themselves socializing with celebrities, only to get in trouble later on career-wise due to their flagrant lack of respect. Even photos of celebrities without you in the shot could be controversial. I was recently visiting a show and one of the crew members showed me photos of the show that he took from his unique backstage location. He said that he wanted to post some on his website, but had to get permission first from the management. That was the right attitude, as the angles and perspectives of his photos were different than those of all the audience members with smart phone cameras. Perhaps the artist or production management did not want to reveal imagery from this other position.

Autographs Are for Fans

If you have ever been on a tour, you probably learned the unwritten rule about the toilets on the tour bus. Another unwritten rule in the entertainment industry is to *never* ask for autographs. You would not ask the audio tech for his autograph, so why would you ask that of the person performing on stage?

Sometimes you might get an autograph as a gift, but you should never ask. I remember when I had a check for a show that was personally signed by Oprah Winfrey. When I went to the bank, the tellers could not believe that I would rather have my money than keep the signed check. I insisted that I wanted to be paid for my work and preferred the money over the signature. The teller reluctantly cashed my check and also provided me with a copy of the check so I could keep the autograph. However, I gave her back the copy, as she was more thrilled about the signature than I was.

Some artists will be very friendly with the staff and may offer to sign items for you. Be sure to allow this to happen naturally and openly. If you manage to get signed memorabilia, don't run out and sell them. They are personal gifts to you that should be cherished.

Keep Your Cool

Remember that celebrities have many people who want to talk with them. If you are at a show and need to have a discussion with the artist, you might have to wait a bit. Sometimes artists will call you into the dressing room before or after the show to discuss the evening's performance. Remember that they have very busy schedules and thus you might need to accommodate their timing needs.

Don't think that just because you are involved with the sound, you have an open invitation to socialize with them. They have many other obligations as well, and while your dialogue is very important, try to remember that they are being pulled in many directions at once. When you do chat with celebrities, keep it professional and always bear in mind that they are at the "top of the food chain" regarding the performance. However, do not let their status compromise your professionalism or the safety of yourself or others. Feel free to inform them when they are wrong or an idea is dangerous or inconceivable.

Of course, we are all human and will find times when we are personally thrilled to meet certain famous people. I will never forget the moment when I shook the hand of astronaut Buzz Aldrin while working with him at a corporate event. Take these moments in stride and remember that your career sometimes awards you with these special moments.

We Are All Celebrities

As I stated at the beginning of this chapter, being famous just means "well-known." Consider too that there are many different types of celebrities. "Famous" is not reserved for those with worldwide public recognition. You might get a chance to work with a well-known sound designer or production designer and feel more butterflies in your stomach than you do when you work with a movie star. Perhaps this sound designer is a bigger celebrity to you than any movie star, and that is okay.

Whether a big-name FOH mixer or a world-renowned pop star, every person should be treated merely as another professional coworker. Sometimes they will become your friends and sometimes they will be a thorn in our side. Either way, famous people are commonly an important part of our lives when we work in the entertainment industry.

28 Healthy Living

"But what if I die?" was the question I heard the designer ask when I told him to go back to the hotel and recover. At some point in our careers, we will experience working with ill people or being sick ourselves during a show situation. Unfortunately, the age-old credo, "the show must go on," often takes precedence over our own wellbeing. Of course, no production is worth dying for, and in the same way we must consider what benefits we really provide to a show when we are feeling under the weather.

Health Matters

Hours of late nights, high stress, poor eating habits, and dirty conditions will lead most humans to some state of illness. Although there are many preventative measures that usually help, illness is bound to eventually happen. I was working a theatrical production during which the designer was not feeling well all through load-in and pre-production. He was overworked, overstressed, hungry, and sleep-deprived.

I decided that enough was enough and suggested that he return to his hotel room to get some rest. I suspected that if he could rest for a few days, he would be in great shape by show time. First, I rounded up the technical team to explain the situation to all of them. Then they helped me to assure the designer that we could cover all that needed to happen in the next 48 hours without him. Only with total moral support from his team did he feel comfortable enough to get some rest.

He actually slept for over 30 hours straight! After some rest and medication, he was able to return for the final rehearsal and the show opening. Of course, we had all pitched in to assist while he was out, and he was pleasantly surprised by the amount of work that had been accomplished without him.

Sometimes you need to help out a coworker and let them know when to throw in the towel. Frequently when a person is sick and overworked, it can be difficult for them to choose their own health as a priority.

Sick as a Dog

When touring, there is usually no backup person ready to fill in the position of mixing a console for a single night. Due to the mixer's particular ear, the

DOI: 10.4324/9781003188957-33

desires of the band, and the operation of the desk, there are very few people who can instantly take over a show at the last minute. Unfortunately, this means that the touring mixer often has to run the show regardless of their condition.

Of course, there are rare times when a person is so sick they cannot continue with a production. A friend of mine was working on a corporate event as a systems engineer. On the first day, the monitor engineer said he was feeling bad and needed to go to his hotel room to rest. The next morning, everyone discovered that the engineer had been so sick that he checked into a local hospital overnight. Suddenly, my friend moved from system engineer to monitor engineer and found himself staring at a console with which he was unfamiliar. However, he quickly took on the tasks and helped the show move forward.

The original monitor engineer made the right choice by deciding that his health was more important than a simple corporate event. He knew that someone on the staff would cover and guarantee that the production continued as normal. In the end the guy healed and the show continued without him. Remember there will always be more shows, but you only have one life.

Staying Healthy

Every bookstore in the world has volumes of books on how you can remain healthy, but the following are some sure-fire tips that can help when working in the industry.

Wash your hands often—You spend all day touching cables, gear, and a sound console, which are often handled by others as well. Washing or sanitizing your hands is imperative, especially before eating anything.

Eat well—Your body can only take a limited amount of cold pizza and burgers at 3 a.m. Fruits and vegetables will help replenish important vitamins.

Get some rest—Sleep rejuvenates our body and mind. Skip going to the bar one night and instead go to bed early. Try to achieve eight hours of sleep.

Vitamins are your best friend—Many people take vitamin C to keep their immune systems strong and vitamin B12 for stamina. Daily multivitamins are helpful too.

Drink lots of water—Our bodies are made of mostly water, so drink as much as you can to keep fresh. Health authorities often recommend eight eight-ounce glasses per day.

Get some fresh air—Sitting in a venue for fifteen hours a day can be both physically and mentally draining. Go outside periodically and sit or walk for 30 minutes to soak up some sun and fresh air.

See a doctor—When you start to feel bad, go to the doctor or ask the production office to call for an onsite medic. It is much better to have a professional diagnosis and treatment than to continue to suffer.

Limit alcohol and drugs—Some recreational usage is acceptable for many (though always remain within the law), but extreme use could lead to health problems, lack of sleep, and addiction.

The Show Will Go On

One key element to remember is that, in most cases, the production is about the act or event, not your individual contribution to the sound. If you have to go to the hospital, the show *will* find a way to continue. With some simple preventative measures, you can usually avoid these situations, but you should also be prepared for when you are sick. When our bodies are not well, we usually do not perform at our best and the production will suffer. Whether a concert, theatrical event, house of worship event, or corporate presentation, all productions *can* continue without your personal contribution. There is no show worth sacrificing your health and life over.

29 Time to Sleep

While working in any portion of the entertainment field, particularly in audio, one will find periods of time during which they do not get enough sleep. Audio professionals are especially susceptible to sleep deprivation as their working duties often need to be scheduled around other production needs and can take many hours. Sleep deprivation can be extremely dangerous as it slows down cognitive functions and impairs most tasks. Audio staff will inevitably be placed in situations requiring less-than-ideal sleep, and it is good to be prepared for these conditions.

Good Sleep

Experts agree that adults should get between seven to nine hours of sleep a night. This rest ensures maximum physical and mental acuity while promoting health and wellness. While I generally maintain this sleep goal when at home, I cannot think of any production period during which I was able to sleep this much on a regular basis.

Unfortunately, many audio processes need to happen when the stage is free of other departments. This generally means working late at night after rehearsals, set builds, photo calls, or other needs. During some productions, you may find it best to work as a "vampire": working during the night and sleeping during the day. Keep in mind that it can be tough to swap your schedule from the norm and then swap it back once rehearsals start. These transitions often cause sleeping difficulties. Anytime you are sleeping during the day, you need to ensure you have the proper environment. Close all curtains, keep the room cool, and stay away from television, phones, computers, and other screens. Put your phone on do-not-disturb and strive to get as many solid hours of sleep as possible. Proper rest should be a top priority in your life.

Bad Sleep

When you get little-to-no sleep, your brain begins to get foggy and will not process actions or the world around you very well. You will find it difficult to speak complete sentences and you may even forget simple tasks that are part

DOI: 10.4324/9781003188957-34

of your regular routines. Mistakes will become more common and your under-standing and reasoning skills will suffer. Working long hours without sleep may appear as if you are accomplishing more, but in reality you are introduc-ing more potential errors and problems into your workflow.

Many productions simply do not allow enough time for a proper sleep. For instance, I was working on a corporate event that had a 6 a.m. lobby call time. We then traveled to the venue, rehearsed some segments, and started the show at 10 a.m. The show finished at 8 p.m.; then we had more rehearsals until midnight. Once this was complete, I needed two more hours of programming to prepare for the headlining band that was scheduled to come the next night. So I returned to my hotel at two-thirty in the morning, only to have another 6 a.m. call time. A similar cycle repeated for four days! As the days continued, I certainly was suffering from lack of sleep and exhaustion. Plus it was not just me; the entire production crew was affected by these crazy required hours. Not only were we exhausted, but we were also all beginning to speak differently and make minor mistakes. Going to catering was like visiting a strange zombie land of catatonic people!

Schedules like the one described above seem to be more and more common as productions get bigger and more complex. In many cases during my career, I have gone multiple continuous days with little-to-no sleep. Concert touring is built around a similar cycle with full days followed by short sleeping time on a bus. In any of these situations, it can be nearly impossible to achieve the recommended eight hours of sleep on a regular basis.

Sleep deprivation may not only be a result of the actual working hours. Many shows combine travel and work for efficiency; however, this rarely results in rested employees. If you have to travel for six hours and then immediately work another ten hours, you will be at your limits as to what you can achieve. Furthermore, many of us steal away sleeping hours on the road to go out for dinner, drinks, sightseeing, or other activities. While it may be entertaining, you will pay the price for your night out when the 6 a.m. call time comes!

I'll Sleep When I Die

When working on a production with limited allotted time for sleep, it is best if you can find some time for a quick nap. With a touring show this is easiest because the bus with your bed is generally not too far away. However, many other productions simply are not set up for napping. Personally, some of the places I have slept include on the floor between the seats in an arena, in my programming chair, on a road case, and in the production office. Studies show that even ten minutes of rest help rejuvenate the brain and body.

If you are fortunate enough to sleep on a plane, this is another opportunity to catch a few Z's. I find that when I am tired I sleep like a baby from the moment the plane starts rolling until we land. However, I know others who suf-fer throughout the entire flight wishing they could sleep just a bit. Sleeping on a plane is a good use of the time, but trust me, it is never equivalent to sleeping on a nice bed (unless you are in premier first class).

Fighting the Deprivation

Lack of sleep is inevitable for most audio personnel. No matter how you plan for the deficiencies, sleep deprivation will take a toll on both your body and your productivity. There are many stimulants that can be used during times of limited rest to help keep you awake and alert. Coffee tends to be the preferred choice among many. The caffeine in this readily available drink will help provide a more alert state of mind and keep your motor running for hours. Tea, soda, and energy drinks each work in a similar manner by delivering caffeine or comparable substances to our brains and helping to stave off the effects of sleep deprivation. Be sure to stop the intake of these drinks a few hours before the end of your workday. Otherwise, you may find yourself lying in bed unable to sleep.

As you continue working without sleep, your body will crave sugary and starchy foods such as candy, chips, and other unhealthy items. These items will help your body survive in its sleep-deprived state, but can have health consequences later on. Weight gain and increased cholesterol levels are to be expected with longer bouts of less-than-optimal sleep. Try to stave off the sugary snacks and instead go for healthy sources of energy boosts such as bananas, apples, and nuts.

There are also energy supplements and drugs which enable people to remain awake for long hours. These are extremely dangerous (and often illegal) as they lead to impaired thinking, poor judgment, and dangerous health risks. Although they may be common in some circles, illegal drugs should never be consumed during working hours as an aid to combat sleep deprivation.

So Very Tired

Sleep is a very important human function—just as important as eating or breathing. Do your best to not allow productions to overwork you to the point where you become sleep deprived. Although long hours are common in the industry, it is up to each individual to take care of themselves and achieve quality rest. Try to nap when you can, eat healthy, and only make sparing use of caffeine and other healthy energy aids. If you must go a night or two with limited sleep, get to bed as soon as possible and make it a priority to sleep for eight or more hours. Remember, you only perform at your best for a production when you are properly rested.

30 When It Stops Being Fun

Working in the live entertainment audio field provides happiness and joy to many. However, there are some unfortunate situations that can emotionally drain you to the point of wanting to quit a gig or audio altogether. Don't let these rare occurrences chase you off, though; instead, use them as opportunities to learn for the future. However, some state of affairs with unforeseen situations could force you to speak up or quit a show.

If your overall experience and mindset are in good shape, you will survive to continue along with your career. It is important to identify a downward spiral as soon as possible, after which you can make informed decisions to respond accordingly. After all, you want to continue to work on productions that fuel your passion for audio.

Going Down Fast

No single event or person has ever caused me to change my attitude towards a gig. When the change occurs it is the result of many factors all coming together against the greater good of the production or directly against me. Usually lack of sleep and improper nutrition play a role as well. When a coworker yells or is rude, remember that they are likely stressed and may not actually mean what they are saying. Take the time to calmly respond to the rant and solve the overarching problem together.

In other cases, there might be personality traits or attitudes that simply do not align with yours. It may be best to keep your head down and complete your job with as little conflict with the opposing person(s) as possible. Thankfully these are rare instances that usually work themselves out with proper communication.

The Emotional Drain

A few years back, I was involved in a production that had many factors come together to knock me down emotionally. The production was a success from the audience's point of view, and everyone involved was happy in

DOI: 10.4324/9781003188957-35

the end. No one was fired and the show went on without a hitch. However, getting to the last day was definitely an uphill climb. In the midst of the turmoil around and within me, I wrote some notes about how I was feeling. It was an unusual feeling for me and I truly was unsure how to deal with it in the moment. Here is a portion of my diary entry that summarizes the state I experienced:

> Right now I feel tired, frustrated, and I feel like I have lost all will to care about anything to do with my work here. I really don't like how I feel in this moment. I truly don't want to be involved anymore. I now believe I am working for the paycheck and not for the passion inside me. That sucks. I feel beaten to a pulp and that my input, experience, and advice have been ignored at every turn. I don't have to be correct . . . I just want the best outcome for the show.

The passion for a great show that usually lives inside of me had been alive and kicking throughout the pre-production period. I worked hard to ensure a top-notch production while collaborating with all of the production staff. Problems kept arising from every angle and anything I did or said was challenged due to politics and budget. Throughout it all, I continued doing my job, as well as assisting many other production positions that were failing.

The very passion that always burns within me had been nearly extinguished when I wrote that passage above. It saddens me to think that I was driven to that low of a level. After the show ended, I took some time to contemplate the occurrences and debrief with others where possible. I also decided not to work with certain members of that production in the future.

Speak Up

Imagine you are busy working on a show or in a venue and you find yourself going down a similarly terrible path. The first thing you should do is calmly speak up with other members of the production staff. It is easy to sit around and complain, but this never solves anything. The best thing to do is to call a production meeting and calmly list the problems at hand. Explain that the production is in trouble and that you have reached the end of your rope. Ask for help in a calm manner and I bet you will be surprised by the results.

Too often people let their emotions drive them and they begin yelling and making others mad. Calmness is the key here. In my previous example, we managed to solve many problems in the meeting I called, in which I explained peacefully exactly what was going on. Other members of the staff were then willing and able to help resolve the complications that had existed for days. My relaxed presentation of the problems helped set the tone for the meeting and encouraged others to refrain from raising their voices.

Watch Out for Murphy

When frustration, sleep deprivation, emotional disruption, and other negative influences strike, you should be on the lookout for Murphy. When we are in these states, it can be difficult to respond appropriately to the unknown situations that occur according to the "Murphy's Law." A FOH engineer who might be adept at dealing with a power outage at FOH could overreact if he is already disappointed with other problems on the show. In a normal emotional state, he would be able to recover accordingly with little impact on the production. However, when he is already drained and beaten, he is likely to make further mistakes as he tries to recover from this new problem.

Be on the lookout for the unexpected and try to respond instead of react. Once you have a negative outlook about a production, suddenly everything you see tends to favor the negative. Remain as optimistic as possible and don't let several small problems appear as one big depression zone.

Be Aware

The real key to remaining happy in the entertainment audio field is to be aware of your own emotional state. There are millions of self-help books on this very subject because humans find this process of self-awareness difficult. Every moment of life cannot be happy and perfect, thus we must learn to analyze situations so that we can restore as much joy as possible to our lives.

In addition to learning audio skills, take the time to work on yourself for the betterment of all around you. Books, podcasts, seminars, therapists, and other resources are all available to help you grow. If you practice becoming aware of your own internal feelings, your newfound insight will establish better relationships with all for your next production.

Once you identify how the production environment is impacting you personally, you can take action to correct it. Many engineers, designers, and other audio staff have made the decision to walk away from a production in order to preserve their own emotional states. Once they realized that the conditions were harmful to their industry passion and internal emotional state, they chose to heal themselves by removing themselves from the negative environment. Their careers then blossomed even further as their internal passion burned stronger than ever. Be on the lookout for joy-draining circumstances and take action as soon as possible to restore your passion.

Part 6

The Travel

When you are working in the entertainment audio business, at some point you will be asked to travel for your work. Even venue employees are sometimes tasked with attending a tradeshow or working an event at another property. More often, freelancers will travel to where the gigs happen, either through flights or on a tour bus (or even a combination of both).

Traveling in the entertainment audio industry runs the gamut from the cheap to the luxurious. Some will start out staying in low-cost motels and sharing

DOI: 10.4324/9781003188957-36

driving duties with the band members in a small van. You might have to share your hotel room with others and eat fast food. However, the average traveling person in the field will stay in decent hotels, travel on buses and/or commercial flights, and devour fresh catering. Overall, the traveling is very reasonable and comfortable.

On the other end of the spectrum, certain sound designers, audio engineers, and crew will earn the opportunity to upgrade their travel experiences. Some will travel on chartered flights, ride in limousines, and stay in five-star hotels. In extreme cases, a few will travel with the artists on private jets, attend exclusive meals and parties, and generally follow the celebrities as a part of their traveling party or entourage.

The majority find traveling in the entertainment audio field very enjoyable. In nearly all cases, the production (or the company that employed you) should pay all your travel expenses. Never should you have to shell out your own money for your job. Be sure to get reimbursed for taxis, ride shares, baggage fees, parking, and other common travel expenses. The following pages detail the very basics you need to know to travel like a pro.

31 Travel Is Work

First and foremost, you need to understand that the very act of traveling is labor and it should always be thought of as such. Spending your time waiting in lines, going through security, sitting on planes, riding to and from the airport, and other travel minutiae occupy a major portion of your valuable time.

Traveling Value

Always negotiate payment for your travel days. You certainly cannot work other gigs while you are getting from A to B, so the current gig should compensate you accordingly. While you may not always earn your full day rate for traveling, at least half is totally acceptable. If you are paid a weekly rate, then ensure that the travel days fall within the paid week or that you are paid for the entire week in which you travel. The paid week should not start after you arrive on site.

You might decide to do some work while on the plane or at the airport. Producing drawings, configuring software, editing music, creating patch sheets, and other computer-related tasks are easy to accomplish during your commute. Remember, however, that the actual travel is your task at the moment. Depending on your contract, you might have been able to charge for those hours if you had done the work at your desk before departing. Once you are getting the travel rate, you likely will not get reimbursed for additional work you perform during the voyage. Of course, if you use the travel time wisely by working on a different production, then perhaps you can "double-dip" by charging one client for your travel time and another for your office time.

Time Is Money

Some customers may ask you to travel and work on the same day. They will book you to fly early in the morning and then expect you to come to the venue at noon for load-in. Or perhaps they ask you to mix the audio for the corporate event and fly out a few hours after the show ends. These timesaving measures will burden you immensely; thus, you should ask for appropriate compensation.

DOI: 10.4324/9781003188957-37

For instance, if during one day you travel for four hours and then work for eight, it should not count as a single twelve-hour day. Instead, you should charge for your travel rate plus a day rate. Let the client simply save on hotel costs, but don't allow yourself to work for less money than you deserve for your time and effort. Reasonable sensibility should always be applied here. If your client asks you to fly all day and then attend a quick production meeting, it is probably not worth asking for additional money to cover both.

When working for a company, as opposed to being a freelancer, you will need to abide by the travel policy of the business. As you are a salaried employee, the company will see your travel time as a part of your job. I still advocate for viewing the journey as work and suggest not multitasking while traveling. You should manage your time accordingly so that you can relax and focus on the travel rather than overwork yourself.

32 Earn the Perks

The travel industry offers many rewards, incentives, and upgrades, and you should grab all that is entitled to you. It is strongly recommended to sign up for as many travel service reward programs as possible. Airline miles, hotel points, rental car plans, and other loyalty programs reward you with perks and benefits that you may not otherwise be allowed to enjoy. I have been very fortunate that, over the past twenty years, I have not had to pay for flights or hotels during my vacations. Instead, I have made use of points and rewards to cover these costs.

Airline Miles

Airlines' frequent flyer programs are the most lucrative for any traveler. Beyond earning points for flights, they also present benefits such as preferred seating, no baggage fees, upgrades, minimal lines, free drinks, and early boarding. Additionally, if your flight is delayed or rebooking is needed, having higher status will certainly aid you in these rough situations. Remaining loyal to one airline will earn you the highest status; therefore, you should try to fly the same airline (or its partners) as much as possible.

This may be easy if you are booking your own travel, but business travel is often booked through a production's travel coordinator. You can always ask for bookings on your preferred airline and many will accommodate as long as the fare difference is not too great. Even if you are booked on another airline than your preferred one, be sure to sign up for their frequent flyer program to guarantee you are earning points on that system too. When flying, the miles are generally applied to the passenger and not the booking party, so make certain that you are earning all that is available to you. Then remember to use these miles the next time you need to book personal travel.

Hotel Points

Hotel points are useful as they can be redeemed for free stays. As with airlines, it helps if you can remain loyal to one hotel brand. This way you will gain the most status and opportunities for upgrades, bonus points, and other perks.

DOI: 10.4324/9781003188957-38

Hotel points are often more difficult to earn because they frequently apply to the payment holder's account. If you are traveling on tour, or if the production staff books the rooms, then it is likely that you will not earn points for your hotel stay. This is because the room rate permits only the booking agent to earn the points on the group package.

However, if you are working for a company that allows you to book hotels on a corporate or personal card, then the points will likely be yours for the taking. Similar to airline miles, be sure to sign up for the point system of every hotel brand in which you happen to stay a night. Keep track of your loyalty status through a smart phone app and sign up for extra bonus point programs when they are offered.

Also remember that any purchases made at the hotel can usually earn points, whether you booked the room or not. If you buy meals, sundries, or hire laundry services, your spending will usually be eligible to receive points. Always provide your membership number to make sure you get credit towards your purchases.

I once was able to unlock expired airline miles because I had previously purchased ice cream at a hotel gift shop while on tour! Thankfully, I had earned three points because I provided my hotel membership card when I bought the ice cream. These points then enabled me to reactivate airline miles due to a partnership between the hotel and airline. You never know when a simple purchase will lead to further perks and benefits.

More Benefits

Rental cars, shared ride services, airport parking, restaurants, and other vendors that you utilize during your travels usually have reward programs. The more you can make use of these (especially when spending per diem or expensed dollars), the more you can take advantage of free or discounted items in your personal life.

In many cases, points and miles can also be redeemed for many different services and products. Sites such as points.com allow you to exchange your loyalty points for cash, services, gift cards, and products. I know of one audio engineer who traded in his airline miles for a new oven in his home!

It's So Easy

Sign up for all the various programs you can find to maximize your reward possibilities. Then store all the membership cards in a convenient location and install the various apps on your smart phone or tablet. Check your status often to confirm you are receiving points and sign up for additional bonus programs when offered. Within no time, you will amass enough points and status to gain further perks and joy from your travels.

33 Air Travel

Flying has become the most common method of transportation around the world for business. It is also often touted as the safest way to get from point to point. However, many people find the entire experience extremely stressful and aggravating. If you know how to properly manage the time, flying can easily become another simple task in your business and personal life. The following outlines some tips to help you have the best flying experience possible.

Pre-Flight Planning

Always select your seat and double check that it is correct at the time of booking. Whether you are booking the flight or it is booked for you, the first thing you should do when receiving your itinerary is to check that the assigned seat is one you actually want to sit in. Usually you can log onto the airline website or app to see a seating chart. Depending on your status, you may be able to select a better seat for no fee (otherwise additional costs will be shown).

Sometimes a travel coordinator will book a flight without selecting seats, and this will default you into a middle seat in the back of the plane. You are the one who will sit in the seat, so take the time to select a seat that you would be happy to plop yourself into. If your preferred seat is not available, keep checking back (especially a few days before the flight) as new seats may open up. On a recent flight, I managed to change to a much better seat the day before my flight departed.

When packing, decide if you are going to carry on your luggage or check a bag. Checking a bag usually will cost additional dollars, but this fee is often reimbursable with the production or company that employs you. Frequent flyer status, credit cards, and other perks may allow you to receive free bag checking as well. The choice to carry on or check may not even be an option if you are hauling tools or other specialty equipment that are not allowed in the passenger compartment of the airplane. Otherwise the choice really is an individual preference.

Personally, I prefer to check my bag unless I am on a direct flight or will be in a hurry upon landing. I find that competing for overhead bin space can be stressful and I also don't like to roll my suitcase around the airport. Regardless

DOI: 10.4324/9781003188957-39

of what you choose, your suitcase should have your contact information on and in it as well as some personal identifiers on the outside. If you can easily distinguish your bag from others, it will reduce confusion for yourself and other passengers.

If you are a frequent traveler, it makes the most sense to have a separate set of toiletries and other common items instead of having to pack and unpack them for every trip. You can just refill the small containers as needed between trips and keep your suitcase ready to go. Learn to efficiently fold your clothing and only take the essentials on your journeys. Multiple shoes and large coats add bulk and weight and should be avoided unless absolutely required. Never pack your essentials, like phone and computer charging cables, in a checked bag. Always carry these in your backpack as you never know when you might need to plug something in.

At the Airport

Arriving at the airport one-and-a-half to two hours before your departure time should provide adequate time to check bags, navigate security, and trek to your gate. Of course, common sense should be applied as some airports or departure times require additional time. For instance, early Monday mornings at all airports are generally busier than during Wednesday evenings. If you are checking bags, it is normally fastest to use the kiosk system to check in instead of waiting for an agent. In many cases, the kiosks are also faster than the priority or first-class lines too.

After checking your bag, you will need to head to security. There are generally three or four different methods of access (and a few others that are out of site). If you are simply a general passenger, you will be in the longest line. Some frequent flyer and credit card perks will enable the use of the priority line, while many also sign up for a pre-screening system such as TSA pre-check (www.tsa.gov/precheck).

With a pre-qualification system, the security line will move faster as passengers have been previously screened (and paid a fee) to reduce the scrutiny required for airport entrance. Because everyone in this line is not removing their shoes, electronics, food, and other items, it tends to move faster. There are also higher-priced services that allow you to cut in line at the checkpoints. The biggest VIPs have secret security areas with little to no wait, but those are normally out of the average person's reality.

Once through security, you will be free to move about in the terminal. It is best to get close to your gate area and confirm that your plane is on time. Food and drinks at the airport are expensive, but you should always get some water. Because you cannot bring liquids through security, you will need to either purchase the water or bring a container to fill. Many airports have bottle-filling stations, which are easy and free to use. Others simply have standard water fountains, but these often dispense foul-tasting water. Additionally, you can usually ask a coffee shop or another vendor to fill your water bottle for no charge using their filtered water.

Unless you have an ailing body, I strongly advise standing and walking prior to your flight. Personally, I find walking in airports a very relaxing and enjoyable activity. This is also why I often check my bag, as I don't want to drag it around as I walk for an hour. With my backpack on my back and earbuds in my ears, I set about to walk the entire expanse of the airport before boarding my flight. About ten minutes before boarding, I will stand in the gate area waiting to board the plane. By the time I am in my seat, I am usually ready to sit down for several hours. I have never understood why some people choose to sit for an hour at the gate before boarding a multi-hour flight, only to sit longer.

On the Plane

Prior to boarding the plane, you should prepare yourself with essential items for your flight. These items will aid your comfort, entertainment, and safety, as well as prepare you for the unexpected. A small backpack or bag with these items can be placed under the seat in front of you for easy access. Plus the majority of these items will also be useful at your work site too.

Always bring water and a snack on every flight, regardless of its length. I was recently on a two-hour flight that included a five-hour delay of sitting on the plane before we took off. Although in this case the airline did pass out water and snacks, often they do not. You should maintain in control of your own food and beverage options and not rely on the airline. Plus, if you happen to have a long flight with a meal service, there are good chances that the food will not be to your liking.

The flight attendants will warn you to fasten your seatbelt anytime you are in your seat. Trust me, you want to do this! It does not have to be tight, but significant turbulence is always possible, and you don't want to be thrown about. Make it a point to stand up for at least five minutes every three hours, and always be cordial to your fellow passengers and flight crew. If something goes wrong on your flight, don't panic. Look around at the flight crew to determine the severity of the disruption and follow all their instructions. Remember that flying is continually proven to be the safest mode of travel. I have been in several emergency landings and thankfully none resulted in major problems.

Flying is a great opportunity to catch up on personal needs such as reading, podcasts, music, and movies. You may be inclined to work during the flight, but often this is difficult due to the limited space and the prying eyes of other passengers. Some people are able to sleep on planes (thankfully I am one of them), and flying is a great time to catch on some sleep. Remember though that unless you are in a first-class lie-flat seat, the sleep you get will not be as restful and rejuvenating as you might expect. Always make the most of your time on the plane so that you have a great experience during your journey.

Table 33.1 On-Plane Essentials

Item	Purpose
Noise-Blocking or Canceling Headphones or Earbuds	Drowns out background noise and allows clear listening to entertainment of your choice.
Ear Plugs	Silences your space when you are trying to sleep or read.
Water	Always bring your own water. Purchase bottles at the airport or use a refillable container.
Snacks	Nuts, protein bars, and other munchies help reduce hunger.
Entertainment Devices	Smart phone, tablet, laptop, electronic reader, and other items help pass the time.
Reading Materials	Books and magazines can provide entertainment and/or learning.
Power Cords, Plugs, and Adaptors	Bring what you need to plug in your devices at the airport and on the plane. Always anticipate needing a charge.
Pen	When travelling internationally, you may be required to fill out landing documents.
Pillow	Some like to bring a travel pillow for improved comfort.
Warm Clothing	Planes are often cold. Have a light jacket or beanie cap in your bag in the event of cold air.
Hand Sanitizer	Always a good idea to keep healthy. Apply before eating anything on the plane.
Cleansing Wipes	Wiping down the area around your seat helps to reduce the spread of bacteria and viruses.
Medicine	Bring essential medications and a standby of pain relievers, anti-acid, medicine for nausea, and other common remedies.
Flash Drives	If you are responsible for maintaining data such as plots or show files, have backups on flash drives with you.
Batteries	Some devices may require a battery change or a backup battery for additional energy when plugs are not available.
Bag Tag Receipts	Always keep the tag that the airline provides when you check your bag. This will be essential if your luggage is missing.
Computer	Never check in your laptop or tablet; always bring these with you on the plane.
Socks	If your shoes do not require socks, you might want a pair when going through security or relaxing on a long flight.
Sleeping Aids	Certain drugs, herbs, and foods will help you get some rest.
Toiletries	It is best to bring some basics with you in case your luggage goes missing or you simply want to freshen up in the lavatory.

Arrivals

When your plane arrives at your destination airport, take the time to wash your hands and use the restroom before collecting your luggage or finding your ride. If you are required to pass through immigration, consider that lines are often long and prepare accordingly. Then once you are in the baggage claim area, stand away from the carousel and watch for your luggage. Keep an eye to ensure that another passenger does not grab your bag by mistake. Once you see your luggage, step forward and grab it. This may seem like common sense, but take a look around at how many people crowd the bag-drop carousel. It can be quite chaotic and this is why I recommend standing back from the mayhem.

Once you collect your luggage, locate the area for your transportation and head out of the airport. You might have to meet up with other production staff to share a ride or look for a runner or a chauffeur holding a sign with your name. Keep any receipts from your journey and be sure to get reimbursed for all your travel expenses.

If your bag has gone missing, don't worry, as it is likely coming in on another flight. Head to the luggage office and provide them your tag receipts (you kept those, right?). They will ask for a description of the bag and any other important information. Often they can tell you when the luggage is arriving and sometimes it may be within the hour. In some instances, it may have even arrived before you on an earlier flight.

Otherwise, provide them your hotel or destination information and they will deliver your luggage as soon as possible (usually later that same evening). If you are on tour though, that can be very difficult as your destination may change every day. In this case, you might find it best to provide the address of a hotel or venue two days ahead to ensure that the bags catch up to you. When your luggage is missing, ask the airline for an amenities kit that will provide you with basic toiletries and perhaps a t-shirt too. Rest assured that very rarely does a properly labeled bag disappear forever.

34 Tour Buses

Many in the live entertainment audio business will earn a living while traveling the world on a tour bus. While the buses may look different in various parts of the world, the basics of this mode of transportation remain the same. Your first time on a tour bus will very much be an outsider experience. You will find seasoned roadies who are extremely comfortable with the lifestyle and they may not share all the aspects of this transportation method right away. Tour bus living is different than other types of travel and it is important to understand its uniqueness.

Bunk Life

Living on a tour bus is comparable to sharing a small apartment with a group of people that you probably would never choose to live with. The total number of people on a bus will vary from tour to tour and can be anywhere from six to twenty. Respect for each other's personal space, food, habits, and routines are essential. Although much of the time on a bus is spent sleeping, there are often long rides during which you will be "living" on the bus. This is when it is crucial to learn to share with your bus mates, while also trying to live your own life.

Typically, there are three main locations on a tour bus: front lounge, bunks, and back lounge. Some parts of the world utilize two-story buses, which further help people to spread out. Your luggage will be stored under the bus in the bays and there may or may not be room on the bus to stash your backpack and other essentials. Sometimes an empty bunk becomes the "junk bunk" where riders can store medium-sized bags.

A small kitchen area provides meals during long travels or even when you want to skip catering. Most tours will supply the bus with essential food and drinks as well as provide a bus-stocking list. You can make requests for specific bus food and the production team will fill the bus accordingly. If you request something you don't want to share, be sure to label it appropriately. Additionally, most productions provide a "warm" meal after the show ends. Pizza, sandwiches, spaghetti, and other meals from local restaurants are often placed on the buses for the staff to enjoy after completing their work.

DOI: 10.4324/9781003188957-40

At the start of a tour, bunk selections and assignments will occur. Sometimes this is an organized process, but at other times, it is a free-for-all. In either case, the general rule of seniority normally applies as the most seasoned staff gets the first pick. Your bunk will become your personal space for the duration of the tour. You may choose to store a few items here, but not too much, as space will be limited and you will want to have room to sleep. The driver will periodically clean your sheets and pillowcase as well as the rest of the bus. For optimal safety, always sleep with your feet towards the front of the bus.

About Number Two

In terms of the restroom, the number one rule on tour buses has to do with "number two." A long-standing decree of touring is to only put liquid waste in the bus toilet. If you feel the urge to do other business, you will need to ask the driver to find a rest stop for you. Although bus technology provides non-smelling systems and easy disposal, this rule is still in effect. Maybe at some point in the future the regulation will be overturned, but until then, don't go number two on a bus.

Shared Space

It is everyone's responsibility on the bus to ensure that the environment is safe and friendly. You should always lock the door and luggage bays behind you when leaving the bus. Secure it as you would your own house. Before inviting friends, fans, family, and others onto the bus for a visit, confirm that your bus mates are okay with strangers entering the shared home. Introduce your visitors and don't allow them to roam freely on the bus, eat the food, or use the restroom. If a person unknown to you enters the bus, ask them whom they are with, and escort them off the bus if they cannot name a bus mate or they lack the proper production credentials.

Generally, people from the same department will travel together on the same bus. This is because the audio staff generally need to arrive and depart around the same time, while the backline crew may finish load-out earlier and can get rolling sooner. However, some tours are now mixing various staff members on the same bus. This way, an entire department will not be affected if germs are spread on a particular bus. This change came about after the COVID-19 pandemic as a way to protect the ability of a tour to continue after someone becomes ill.

Sometimes you might share the bus with members of the band, which may sound exciting, but generally this is not the case. When you are on the band bus, you don't have a chance to remove yourself from your work. You may wish to relax after a show and not continue discussing their thoughts and requests. Furthermore, the artists often have their own needs and desires that may trump yours, simply because of their status.

Certain tours will have buses that cater to specific segments of the crew. For instance, there may be a smoking bus or a sober bus. These designations help those with specific habits maintain their lifestyles while not causing anguish for others. In other situations, smoking may be allowed only in the back lounge. Some people even partake in drug use on buses with feelings of immunity. Tour buses are not sanctuaries and are open to searches by law enforcement and border patrols. Furthermore, underage drinking, illegal drug usage, and other offenses are usually not tolerated on buses. These actions can result in termination and/or prosecution.

Living on a tour bus can be an enjoyable experience as long as everyone on the bus gets along and has respect for each other. If you are having difficulties with specific people on your bus, you can ask the production manager for reassignment. Quite often though, great friendships and camaraderie are formed on buses. Even with great bus companionships, you will find hotel days filled with a new appreciation for the comforts of privacy and aloneness. Tour bus travel is not for everyone, but it can also become a wonderful way of life that many enjoy throughout their careers.

35 Hotel Living

Staying in hotels is a regular occurrence for many in the industry. Personally, I have averaged over 100 nights a year for way too many years of my life. While staying in a hotel may seem straightforward, there are some things you need to be aware of. First, the production will generally book your hotel and will have already paid for the room. You will be responsible for any further incidental charges to the room. Be sure to always pay your hotel bill and settle with the front desk before your departure lobby call.

When production books your hotel, have them provide you with the booking confirmation number so you can check in with ease. This way, if there is no record under your name, the front desk can look up your stay with the confirmation number. When traveling on tour, the production manager will usually check in everyone at once and then distribute the room keys upon arrival. You may receive your key as you enter the hotel or while still on your bus. Either way, make sure you collect your key as soon as it becomes available. Remember, the production manager wants to get to their room quickly too.

In the Room

Fairly often, audio work requires working through the night and sleeping during the day. When your schedule flips from the norm, you need to prepare accordingly. First, bring some gaff tape or binder clips to your room to help seal the curtains. Hotels are notorious for curtains that do not completely seal out sunlight. Another trick is to use the pants hangers with clips that the hotel provides to seal your curtains shut (see Figure 35.1).

Always place the "do not disturb" sign on your door and alert the front desk to the fact that you are sleeping during the day. This way, they will not call to check if you want the room cleaned or disrupt you in other ways. Soft earplugs are helpful too as hotels often perform maintenance during the day, which could further hinder your sleeping plans.

Hotel rooms may seem like a secure location for your belongings, but always use vigilance as with any other location. Keep the door locked and place the do not disturb sign on the door (even when you are in the room). If you need additional towels or a cleaning, just call the front desk. I find that, in most cases,

DOI: 10.4324/9781003188957-41

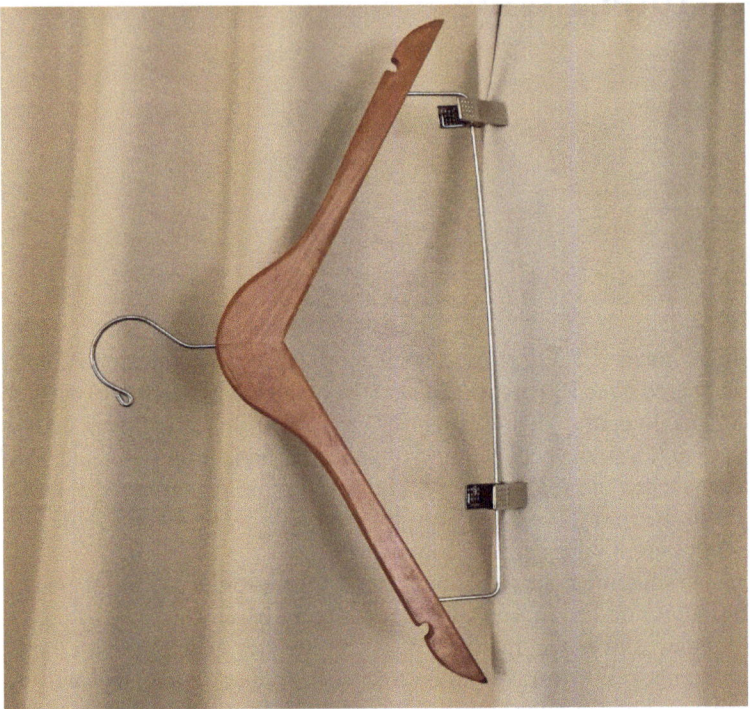

Figure 35.1 Hangers with Clips are Great for Closing Curtains

I am only in a hotel for a few days and do not need staff entering or cleaning my room. Make use of the hotel safe and hide your laptop and other valuables when you are not in the room. Don't forget to collect all your belongings and double-check the room prior to checking out.

Sleep Well

Remember that the main goal of your hotel stay is to get a good night's sleep. You can always ask to move to a different room or floor if you cannot get comfortable in your assigned room. When you check in, ask for a room with a king size bed, as this will likely result in a larger, more comfortable room. Actual room type upgrades generally will cost you money, unless you have loyalty status or points to use. Don't hesitate to ask for the things you need to ensure good rest. For instance, if you have allergies you can often ask for hypoallergenic linens or rooms sanitized with natural cleaners.

As mentioned previously, always sign up for hotel loyalty programs so that you can earn perks, points, and upgrades. With many hotel chains, simply being a member provides you with free Internet, breakfast options, bottled water, and

Table 35.1 Helpful Tips for Hotel Stays

Item	Tip
TV Remote	Many find that the TV remote can be full of germs. Wipe it down with a cleansing cloth or place it in a plastic bag. Some hotels allow TV control via your smart phone so you can skip touching the remote altogether.
Earplugs	Sometimes hotels are noisy. Soft, disposable earplugs will help you achieve a quiet night's sleep.
Hair Dryer	Always point the dryer away from your head when you are first activating it. This way, any dust or hair inside will blow away from you.
Lock Doors and Windows	Keep your doors and windows locked at all times. Double check that doors close securely when you leave the room.
Report Problems	Let the front desk know about any problems you might experience at the hotel. Often they will compensate you with loyalty points or vouchers for the restaurant or the gift shop.
Hide Valuables	Do not leave important items out in the open. Place them in the safe or behind chairs or curtains. Remember to claim your stashed items before checking out!
Override Thermostats	Many websites explain how to overrule hotel thermostats to achieve better temperatures and disable motion detectors.
Do Not Disturb Sign	Use this sign at all times for security and to keep unwanted hotel staff from your room.
Cancel Housekeeping	Many hotels provide bonus points or extra perks if you call and cancel the daily housekeeping.
Spare Key	Some rooms require a key placed in a master switch to enable the power in the room. Ask for a spare key to insert in this slot so that the power (and AC) remains on at all times.
Extra Items	Forgot something, need another small bottle of lotion, or want a late night snack? Ask for what you need as hotels are great at providing hospitality for their guests.
Late Checkout	If you know you are leaving after 10 a.m., ask for a late checkout at no charge. If you must leave your room, consider checking your bags at the bell desk until you depart the hotel.
Ask the Concierge	Hotel staff often know about the local area and can offer discounts or advice about restaurants, walking areas, laundromats, and other sites.
Use the Gym	If you like to exercise, make use of the provided hotel amenities.
Bring a Long Cable	Travel with a six-foot or longer charging cable for your phone so that you can place it where you prefer.
In-Room Clock	Always check that the time is correct on the clock in the room. Also ensure that the alarm is turned off.
Use Your Own Alarm	Use your own travel clock or smart phone as a wake up alarm. Never trust the hotel clock or wake up calls.
Wear Sandals or Slippers	Hotel rooms are notorious for spreading foot fungus. Bring your own slippers for the room or stand on towels when not wearing your shoes.

other perks. Utilize the hotel app on your phone for additional features such as room selection, digital key, and hotel information. It is also helpful to set up a personal profile that details the type of room and location that you prefer. For instance, you can designate whether you like high or low floors and whether you wish to be near or far from the elevators.

Hotel stays are daunting for some, but routine for many others. Make the most of every stay by envisioning the room as a sanctuary for your sleep. With a goal of achieving the best rest possible, you can enjoy the hotel and take care of yourself. Both the production and your body will be grateful that you achieved some sleep.

Part 7

The Interviews

A career in any industry has many twists and turns through various positions, opportunities, and experiences. People have their own goals and insights, which drive them to very different paths throughout their employment journeys.

I selected six key people in the live entertainment audio field to interview and profile about their specific audio career paths. Each has a tremendously interesting story to share about how they have achieved success. They all also continue to progress forward with only slight thoughts about retirement. There are millions of stories in the industry, and these six are but a few examples to inspire you within your own career.

DOI: 10.4324/9781003188957-42

36 Jim Yakabuski

Director of Audio Projects for Solotech-US Live Productions

Figure 36.1 Jim Yakabuski

Jim Yakabuski's audio career now spans over forty years and he is no closer to seeking out retirement than he was at any other time in his profession. Looking back to his humble beginnings as a young Canadian bar-band mixer, he is pleasantly surprised to have been able to move to the United States and eventually mix some of the biggest bands in the world.

> When I first started, my aspirations were big, but I thought of success back then as just having the chance to mix in an arena someday. To mix in front of 80,000 people in a huge stadium was even more than I could have dreamed and wished for all those years ago.

Jim's career began in 1981 when he was mixing for bar bands in Vancouver, BC, Canada. He readily admits that he was rather untalented at the beginning but very focused on improving his mixing skills. He would read as much as he could find on live sound and experimented every chance he got until he slowly improved. He also asked every audio engineer who was better than him for help, and often they shared their time and knowledge, showing him tips and techniques that helped immensely. While he loved what he was doing, he also knew that he wanted more.

DOI: 10.4324/9781003188957-43

He spent a very short time thinking he wanted to be a studio recording engineer. As a huge Queen fan he listened to their albums for inspiration and read some audio books; then in 1980, he enrolled in the Columbia School of Broadcasting's "Introduction to Recording" class. Jim was about six months in when he first ran monitors for his friend's audition as a singer for a local band.

After a few years of working as a freelancer mixing in the clubs and attending concerts in arenas and big theaters, he was ready to take the next step of becoming a staff engineer at a big Canadian audio company. He researched and discovered that four or five companies were providing audio for most of the concert touring in Canada. He then wrote letters to each one with a brief list of his limited accomplishments with hopes that he might get a chance to work with one of them. Unfortunately he heard nothing back—not one single reply.

> I was feeling very discouraged with the lack of responses when I met a waitress in a bar one night wearing a dB Sound t-shirt. When I asked her about her shirt she mentioned that she was dating an owner of the company and she'd be glad to let him know I was looking for a job with a big sound company. Next thing you know I was on my way to Milwaukee for Summerfest as an audio patch person on the Rock Stage.

The Summerfest opportunity really opened the door for Jim, and not long after, he was on his first tour setting up the PA, patching the stage, and mixing monitors for the opening acts. Then in 1990, after a few tours working as a tech, he was offered an opportunity to mix monitors for Aerosmith. After that tour ended, he mixed monitors for Poison, and by the end of 1991 he was the monitor engineer for Van Halen. Eventually Jim moved to FOH on Van Halen and his unbelievable dream career was in full swing.

One moment Jim admits he will never forget occurred a couple weeks into mixing FOH for Van Halen in 1983.

> I was given a "trial period" and after the band heard good things from friends and colleagues about the sound, I finally started relaxing and settling in. We were at Alpine Valley in Wisconsin, an amazing outdoor venue on the side of a ski hill. The crowd was huge and crawled up the mountain slope. At one point during the show the stage "blinders" came on and lit up the crowd. I turned around and all I could see was people as far as the eye could see. I took a moment and thought, "Holy crap, I'm mixing Van Halen!" I have to give a huge shout out to the Van Halen brothers who let a young monitor engineer earn his FOH wings mixing one of the greatest rock bands of all time.

As Jim's FOH mixing career continued, it allowed him to work with quite an impressive list of bands over the years: Extreme, Whitesnake, Matchbox Twenty, Avril Lavigne Journey, and Gwen Stefani, as well as other artists of different genres such as Julio Iglesias, Engelbert Humperdinck, and Luis Miguel.

Figure 36.2 Jim Yakabuski's Career Path

The life of touring is not always easy nor guaranteed and Jim recounts a time when he came home from a Van Halen tour on top of the world. At the time he did not have another tour lined up. Then the weeks at home turned to months without another mixing opportunity (or any work) and he started to run low on funds. He ended up taking a job at a friend's used car sales lot.

> I was happy to do it, but I remember a day in August in Ft. Myers, FL on my hands and knees washing a car and sweating profusely. I had to chuckle at how my amazing rock 'n roll life had slipped a notch or two.

Fortunately a call came not long after and he was back out on the road mixing for another band. A few years later, he received a call to mix a solo tour for one of his favorite singers. As he had recently committed to another artist's tour and just completed rehearsals, it would have been unethical to jump ship for this other tour. Jim politely declined and let the artist know he would be happy to mix for him on a future tour.

> I was surprised when he tried to talk me into it but I reminded him that he wouldn't appreciate me doing that to him if I was his FOH engineer. That ended the conversation. I don't regret doing the right thing, I just regret missing the chance to work with one of my heroes.

After a long career on the road with various bands, Jim was presented with a new opportunity to work with Solotech as the Director of Audio Projects for its U.S. Live Productions division. However the initial offer came in early 2020, just before the industry stopped due to COVID-19. This meant that the job with Solotech was delayed until 2021 when the industry came back with great gusto. He now happily helps send gear and techs out for countless tours and productions.

Jim is ever grateful for learning along the way and gives credit to Harry Witz and Bruce "Slim" Judd from dB Sound for their teachings when he first came to the United States. Slim taught him a great deal about mixing wedge and side fill monitors in big spaces like arenas. He also admits that he learned tons from each and every audio engineer he toured with as well as the folks in the bar bands that would take the time to teach him things.

While he is appreciative for the wisdom imparted to him during his rise, he is also quick to share his knowledge with others who are along the same path. Some of his best advice is to remember that less is more.

> Learn how to mix "analog" style while you grow as a live audio engineer. Work on tuning a PA system to sound natural, and blend all the speaker zones to sound like one big stereo. Then take inputs into the console without messing with them too much. Get each input to sound good without adding a bunch of processing right off the bat.

Jim plans to continue working for a few more years and looks forward to a time when he can retire and focus on golf and silence.

Time will tell on the "when," but the "what" definitely involves being in the quietest places on earth. And one more thing: road life is hard. It's hard on your family and hard on you. There's no more rewarding job than mixing live audio, but make sure you've thought it through before running off to join the circus.

37 Big Mick Hughes

Front of House Sound Engineer

Figure 37.1 Big Mick

Big Mick is proud to exclaim that he has had a fantastic career mixing sound, yet never worked a day in his life. He absolutely loves what he does and is thrilled that he has been able to earn a great living by following his passion to work with sound.

Best known for being Metallica's FOH mixer for over 37 years, Big Mick's professional career started 1976 when he began working in clubs. He has always had an interest in entertainment since he was young. Growing up in Birmingham, England, he thought he wanted to be a photographer or a cameraman for the BBC; but by the time he was 15 years old, he found that he loved turning the knobs on the small sound mixer at his school. He also enjoyed the art of lighting and did some of that in school too. Just out of high school, Mick had a great opportunity presented to him.

As fate would have it, Big Mick had a cousin named Brian "Bruno" Stapenhill who was a founding member of a new (at the time) heavy metal band called Judas Priest. Bruno asked Mick to come shoot photos of the band during a few gigs and this began Mick's education in live music.

As I traveled with the band, I came to know Bruno more and more. And he was a prolific musician. This was the early 1970s and I got educated in the

DOI: 10.4324/9781003188957-44

school of thought for the way music was for live bands. Bruno was instrumental for me because he was such a consummate musician and had such great ideas that I learned an awful lot from him just from his appreciation of music and what things should sound like.

During this time Mick needed to earn some money, so he also took a "regular" job working at the British Steel Corporation as an electrical engineer. He even went to college during this time to be qualified in electrical engineering. He would travel with the band taking pictures during the night and then get dropped off at work the next morning. In fact, he admits that he was not the most reliable employee as he often missed work due to band gigs being too far away.

As Big Mick learned more about electronics, he found himself useful for other bands too as he could also fix their equipment. He began travelling with other small bands as a technician of sorts. At the time, most bands travelled with their own PA and did not rent gear from a sound company. With one of these bands called Quartz (also from Birmingham), Mick suggested that they rent out their van and gear when the band was not doing gigs. Mick took the lead on these rentals and this was going very well when he met up with Kevan Wilkins, who was renting out lighting equipment to bands. They found that they were often renting equipment to the same shows. Mick also frequently offered his electrical knowledge and service to help Kevan with his lighting equipment.

They became friends, and one day in their early twenties, they decided to start their own lighting company. They teamed up with another friend named Alan Whittaker to form a lighting rental company called Concert Lighting. (Alan Whittaker was a founding member of Light and Sound Design [LSD], which coincidentally started when they provided lighting for Judas Priest).

As Big Mick's lighting company grew, they also started working with Thomas Engineering, manufacturing custom ray-light reflectors for Thomas Engineering's PAR cans. Over the years, they sold thousands of reflectors and provided lighting for various bands.

At the same time as I was doing the lighting company thing, I kind of missed the sound side of it. I still wanted to go and do audio. So I met a guy called Bob Doyle who owned a company called Tex-Serve in Birmingham. So I began freelancing with Tex-Serve and met many established audio engineers and began to learn more about mixing. (Bob eventually went on to run Midas and then later started DiGiCo.)

Throughout this, I never had any formal teaching about sound as such. At that time, you sort of learned on the fly. You learned how to mix by being the guy willing to turn the knob all the way one way and all the way the other way to see what would happen and then work out which part was the good thing to do. And then it was a case of hoping your appreciation as to what tones it should be was correct. That was what it was all about.

In 1980, Big Mick was working with a well-known punk band called GBH. This was when he was able to tour the United States for the first time, albeit in a small van with stays at fleabag motels. He was very enlightened by his personal discovery of the United States and learning how other parts of the world worked. He continued mixing for GBH on and off for quite a long time and enjoyed the experience and money.

Next began an interesting chain of events that led to his eventual working with Metallica. While at home between GBH tours, he received a call from Bob at Tech Service saying he needed a mix engineer for a one-off show for a band called The Armoury Show. Bob told him that it was an important gig at the Aston University in Birmingham, as the band's manager was a very important guy named Peter Mench who also managed Def Leppard at the time.

> I knew I had to be nice to the guy as he was an American and I had only limited experience with Americans from the GBH tours. During the show the PA turned off and I had to locate the drunk PA guy at the bar and get him to turn it back on. I thought the show was a disaster, but the band did sound pretty good when the PA was on.
>
> At the end of the night, I went to Peter Mench to get my 50 quid and he told me that it went well and he liked how the band sounded. I was about to take my money and head home when Peter said, "Are you ready?" I said, "Ready for what?" He said to go to Glasgow (which is over 400 miles away). I said no, that I was told this is a one-off gig and I am heading home. Peter then said, "Well it's a tour, so what do you want to do?"
>
> I told him I had my motorbike outside and no clothes to change into. Peter then said, "You have a choice then don't you? You can go home or go on tour. Either way the band is going on to Glasgow."
>
> So I took a moment to think about it. I was 23 years old and decided to take my bike home and have a mate bring me back to the gig so I could go on tour with The Armoury Show.

He worked on and off with them for about two years. They had been signed by Q Prime and were doing large shows, tours, and festivals. Eventually the band's direction changed and they fell out of favor. Q Prime decided to move on and dropped them to focus on new talent.

> Peter Mench told me they were going to drop The Armoury Show and asked if I want to work for them with another band. They had a new band called Metallica and I asked, "What is Metallica?" Peter said, "If you take the gig you will find out." So in November 1984, Pete Russell and I provided a Tex-Serve PA and monitors for Metallica's European Ride the Lightning Tour.
>
> I was just the guy from the PA company, I wasn't actually the band's engineer at that time. But I never really knew what that entailed at the time either; I just went out and did the job as asked for by the management. After

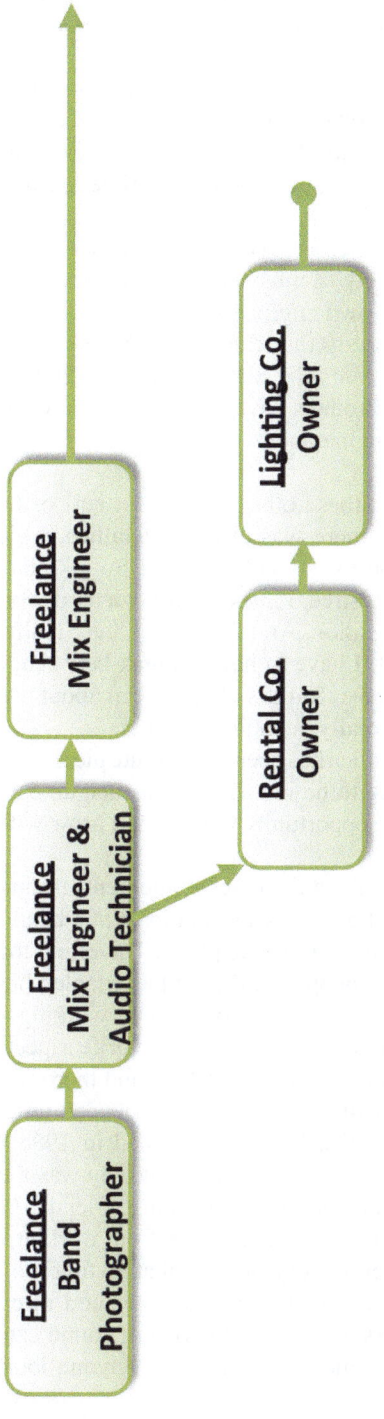

Figure 37.2 Big Mick's Career Path

the European tour the band asked me to be their engineer. In January 1985 I started with the Ride the Lightning tour in America and off we went.

At the same time, Big Mick was still working in his lighting company, which had expanded to include rehearsal rooms. He realized that running a lighting company was more "nuts and bolts" and less creative than what he was getting from running sound. So he decided it was time to leave lighting and focus solely on sound.

Throughout his mixing career, Big Mick has worked with many different bands across a wide variety of musical genres. Back in the 70s and 80s, many bands did not have their own engineers, so he often mixed for bands as a per-gig hire or as the sound guy that came with the PA system. He also put his hands into management for a bit with a band called The Wild Hearts, who was signed with Warner Brothers. Over the years he has continued as Metallica's engineer while also doing many other things.

> There have been all these other things that sort of happened in parallel with Metallica. But there was always Metallica, which has been a very comfortable position for my entire career. So it kind of allowed me to be a little bit more fancy-free, I guess. I'm not a hard-working person by any stretch of the imagination. And I've never, ever really looked to go from one to tour the next. I have to have a break between. I have to take touring in small bite pieces. But it's always been about the way things sound. That's all I've ever really been interested in.
>
> Engineering Metallica has been an absolute pleasure. Not many engineers are given the chance to be involved from such an early stage in a band's career and given the opportunity to learn and grow with them.

Big Mick looks back fondly at the magical moments throughout his career. He loves to tell stories about eating beans and toast in a band's dressing room, or about his first time going on a private plane. Travelling the world and experiencing different cultures, people, and food has enabled him to grow as a person. The camaraderie between the various groups working together and sharing their passion to entertain has always been a favorite aspect of his career.

Of course it was not all happy and perfect, and there were periods in which he was concerned about his next gig or experienced unfortunate events. For instance, when the Metallica band bus crashed in 1986, killing bass player Cliff Burton, Mick was deeply saddened. Not only was the future of the band in jeopardy, but Cliff and Mick were also close friends. As horrific as it was, Big Mick wasn't going to let it dissuade him from working on the road. He and the band worked together to carry on after that awful time.

Big Mick has become a legendary engineer and often teaches his skills through various seminars and talks. His love of sound can be found in all his work and he offers the following advice for anyone looking for a career in sound:

When it comes to sound you have to have your own mental picture as to what each thing should sound like in order to achieve a good final mix. This picture/opinion will get easier to achieve as your skills of operating the equipment improves, but you have to have the picture in your head first and foremost. An engineer with no picture/opinion has nothing to aim for. The picture will change dependent on the project/artists you are mixing. There is help with this picture if the project/artists has already been recorded. You can at least get a view of someone else's picture and the artists too.

Looking ahead, Big Mick has no plans to retire and hopes to continue mixing for Metallica and other artists for many more years. Although the industry has changed immeasurably since he started, he is still having a ball doing what he loves. And as long as he never sees it as work, he will continue to be found behind a console mixing live sound.

38 Jessica Paz

Tony Award Winning Sound Designer

Figure 38.1 Jessica Paz

It is not often that someone starts out aiming for a career in law enforcement and instead turns to the world of production sound design, but that is a true story for Jessica Paz. She originally went to Jay College of Criminal Justice to study forensic psychology. Her aspirations were leading her to an eventual life as an FBI agent. But her passion for sound took her down a more rewarding and exciting career path and she has never looked back at her original desires.

After two years of studying forensics, Jessica decided to take some time off and pursue other interests. Various jobs led her to a friend who was in the cast of the *Rocky Horror Picture Show*. She found that she loved the show and wanted to attend as often as possible, but the tickets were expensive. She decided that it would be less costly for her to volunteer at the theater and see the show for free instead of buying tickets each week. Her first gig was running the fog machine for the show. Over time she took on other positions and eventually was offered a job to stage-manage several shows.

The community-based theater group did not have much money and the production values were low, but Jessica had found her calling in the theater. During one show, she was aware that the sound was particularly bad with a large amount of feedback.

DOI: 10.4324/9781003188957-45

I wanted it to sound better. So I asked a friend of mine how to do a sound check. He basically walked me through how to bring people out on stage and EQ their mics. One night before the show I did a sound check and we finally had a show without any feedback. I then thought, "Oh, wow, that's interesting. That was fun." I started actually mixing the show, because for some reason on this particular show as the stage manager I was also running the lights while the lighting designer was mixing the sound. I traded positions with the LD and enjoyed taking over the sound.

Jessica found a new passion in mixing sound and a friend recommended her to sub in as an A1 at the CM Performing Arts Center in Suffolk County, New York. This led to a full-time position in which she was able to hone her skills and learn from visiting sound designers. Eventually she was ready to design a show of her own and she asked the artistic director if she could design the upcoming production of Smokey Joe's Café. They offered her the job and she became not only their resident mixer, but also their resident sound designer.

After a year and half working at the CM Performing Arts Center, Jessica saw an ad for a sound operator position for an off-Broadway show. The show was *All The Bad Things* at the Public Theatre and sound designer Rob Kaplowitz hired her for the position. She and Rob found that they worked well together and Rob asked her to be his assistant on his next show. This turned into a long-term working relationship and they worked together on countless productions around the world.

Eventually Rob was offered a new show titled *Fela!* and once again he asked Jessica to be a part of his team. Originally it was simply a workshop show and Jessica mixed it for the five-week run. The show continued on to an off-Broadway production in which she moved up to the assistant sound designer role. Then after the show closed, several amazing things happened all at once.

I was visiting Rob in the hospital to meet his newborn baby. I clearly remember that as I held baby Nile in my arms Rob said that *Fela!* was going to go to Broadway, and that he would like me to be the associate designer. *Fela!* had a long life on Broadway and then went on to the National Theatre in London, European and U.S. tours, and even a production in Nigeria.

After a great run with *Fela!*, Jessica moved her focus away from theater for a bit. Although she still worked with a few theater productions, she found a new joy in mixing live bands at various venues in the New York area. She also took a job at *Sleep No More* as the primary audio mixer for the live bands in its bar area. A unique immersive theatrical piece, *Sleep No More* had several different live sound locations, such as a bar, a rooftop, a restaurant, and a patio. She was able to mix for many live bands that came through on a regular basis as well as for lavish private parties.

Shortly after leaving *Sleep No More*, she was reconnected with Rob Kaplowitz who was back in New York City doing another show at the Public Theatre.

He hired her as the A2 and she worked backstage with the actors and their microphones.

> From there I met the audio supervisor who then put my name on a list for an assistant job at the Delacorte Theater for Shakespeare in the Park. I got hired for that. And that's how I met sound designer Mark Menard. I was the assistant at the Delacorte my first year there. Mark then asked me to be his assistant on the Broadway show *Disaster! The Musical*. I believe that was when I asked him to bump me up to associate and he agreed. I then kept the title of associate for my second year at the Delacorte. During my third year there, a few months before we would have gone into the planning stages for the season, Mark unexpectedly passed away; which was very sad and disappointing. The Public Theater offered me the design position in his stead. And that's how I became the sound designer at the Delacorte. And I am still the designer there now going into my fourth year.

From her work at the Delacorte, Jessica went on to meet many more people on Broadway, including sound designer Nevin Steinberg. They worked together on several Broadway shows including *Dear Evan Hanson* and *Bandstand*. On an early production of *Hadestown* in Canada, Nevin was unavailable and sent Jessica is his place. Next came the opportunity to codesign *Hadestown* with Nevin in London at the National Theatre. This naturally led to a Broadway production, which swept the Tony Awards in 2019 with eight wins including best sound design.

Jessica had an intuition and desire early in her career to be the first woman to win a Tony award for best sound design of a musical. This focus continuously drove her to work hard and even helped her come back to theater after a period away focusing on live music. Even still, actually achieving this goal came as a pleasant surprise to her.

One of her favorite audio moments came while travelling with *Fela!* in Nigeria. In addition to the theatrical production, a special concert version of the show was being held at the actual Afrika Shrine venue in Nigeria that the show is written about.

> I got to mix that and I will never forget the exact moment when the show started. I saw our lead actor walk out on stage and I felt more awake than I had ever felt before. I was in awe that we were performing a show about *Fela!* at *the* Afrika Shrine in Nigeria; it was just a pretty bananas moment.

Jessica offers the following advice for anyone working in live entertainment audio: Keep showing up. She applies this motto to all aspects of her work and finds that it has guided her well throughout her career.

> Keep showing up in the room, keep showing up to the jobs, and keep showing up to the people around you. Most importantly, keep showing up for yourself and for your art.

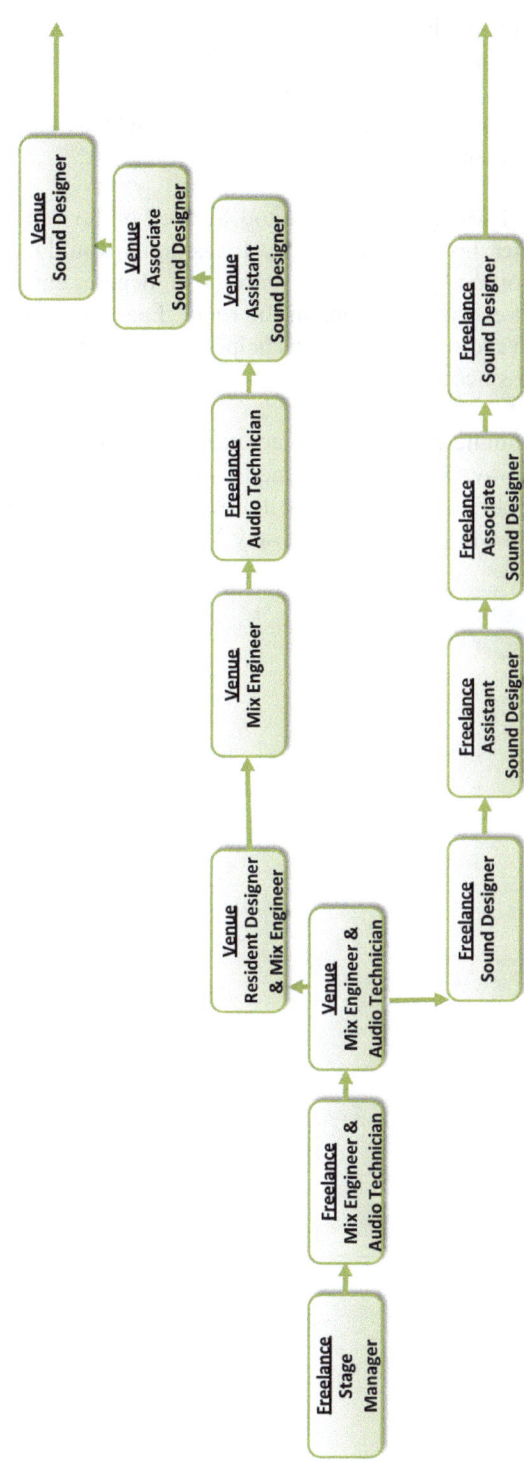

Figure 38.2 Jessica Paz's Career Path

She also recounts the advice given to her by Nevin during a show, when she was having a particularly hard time. He said that every day when he puts his hand on the door to enter the theater he reminds himself to ask how he can help. She thought about this and immediately applied it to her problem (as well as to many other occasions since).

> I realized that I was getting a little too focused on how to make what I was doing better, instead of how what I was doing would make the show better. I just got too hyper-focused on my own department as opposed to thinking about everyone's needs in the room. Learning to ask how I can help before I walk into the room has helped me immensely through many situations ever since.

Jessica plans to continue working for many years to come and sees retirement at least ten years away. Although retirement is just a word to her, because in the future, you might find her renovating houses as a second career. Until then, she will continue to work in theater and stay involved with all types of audio productions.

39 Jim Ragus
Audio System Engineer and Crew Chief

Figure 39.1 Jim Ragus

Some forty years ago, Jim Ragus was playing guitar in bands during high school. While his bands never made it out of the garage, some of his friends had bands that went on to play some real gigs. Jim's career started unexpectedly when he began helping them with their audio needs. In fact, he did not realize that it was even work as he was just hanging out asking questions and learning by osmosis.

> I've always been pretty good at putting things together, so it came rather naturally to me. In retrospect it took a long time to realize that what I was doing was a career.

Jim's first introduction to the professional world of sound occurred a number of years into helping his friend's bands in the regional circuits. As they became more successful, they were finally able to hire a company to provide the sound system. It was then that he realized that such companies and jobs even existed.

Eventually a small sound company based in Austin, Texas saw potential in Jim and started offering him freelance work. Oddly enough, through one of these gigs, Jim met an artist named Ron Loccarini who would have a huge impact on his career. They became friends, and after Ron's band broke up, Ron

DOI: 10.4324/9781003188957-46

decided to change his focus. In a strange twist caused by the debt racked up by his band, Ron became the owner of a business called Titan Sound. He asked Jim to help with this completely new sound company.

> Ron and I actually lived together during his building up the business to a major regional touring company. My time with Ron and seeing that company from the ground up was probably my biggest push in the direction of real professional audio. In fact, I consider this my first real audio job.

Jim was also working freelance throughout this time, mixing regional bands and taking whatever gigs came his way. Although he was living hand-to-mouth, he was actually doing well. Eventually Ron sold his company to M.P. Productions in Little Rock, Arkansas, and they offered Jim a full-time job. He packed up his wife and family and moved into a new role as a system engineer and crew chief.

> My mantra became: "Setup the rig, get things done, and bring home the money." This has always served me well and I am always surprised that I get paid to set up the coolest stereos on earth!

After some time with M.P. Productions, Jim took a job at Showco in Dallas and again relocated his family. Five years into his time with Showco, Clair Brothers acquired the company and Jim continues to work for them to this day. The primary part of his job is to tour with various major bands as a systems engineer.

> The pinnacle of my production career has been touring with The Rolling Stones. I saw them in 1972 as a wide-eyed pup and getting to tour with them was very full-circle for me.

Jim does find it frustrating that working in the business has changed him from the music fan he once was. He finds it very difficult to set his analytical soundman side aside and simply be the naïve music listener he once was.

When asked about retirement, Jim says that the shutdown during the COVID-19 pandemic made it clear to him that he will continue to produce and manipulate sound waves as long as he is physically able to do so.

As for advice, Jim states it succinctly:

> Keep it simple. Our business is complex by its nature so any chance to simplify is a good thing. Point the speakers at the people that paid the money and turn it up!

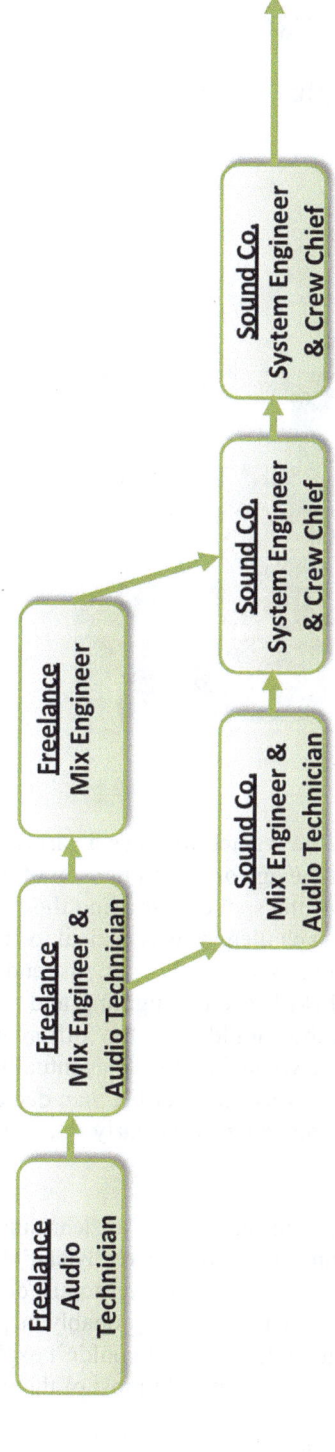

Figure 39.2 Jim Ragus's Career Path

40 Kelly Epperson

Audio Designer & Mixer

Figure 40.1 Kelly Epperson

Kelly Epperson knew what he wanted to do as a career since he started running sound in his church at the age of 13. As the son of a minister, he learned very early the importance of the spoken word and the fact that having things work with minimal technical challenges was critical to engaging an audience. However he was also open to exploring his options, and along the way, tried his hand at other technical skills such as lighting and video. He always had an internal philosophy that he should try a few different things so he could determine what it was that he wanted to do. He eventually realized that deciding what he didn't want to do was a shorter list than deciding what he wanted to do; thus, he was able to recognize very early that sound was a natural fit for him.

> With my dad being a minister, I grew up learning about sound in the church. And being volunteer-based, they really can't turn you down if they want the help. So I found a lot of great ways to just do what I enjoyed. Plus I saw it as an opportunity where I could probably be proficient with sound because I just seemed to understand it. I couldn't explain why I was doing what I did, but luckily I just got it right most of the time.

DOI: 10.4324/9781003188957-47

While volunteering to help with the sound for services at the church, he also found opportunities to run sound for weddings and musicals that the church produced. The musicals also provided him a creative outlet to expand beyond sound into lighting, pyro, lasers, set design, and more. Kelly also became involved in theater and AV in his high school, with a focus on sound. During this time he often rented various audio technologies to assist with the assorted productions. He enjoyed the challenges of learning to operate new equipment and applying the knowledge to shows. By his late teens he had established a relationship with a local sound company called Bridgewater Sound in Harvey, Illinois.

After high school, he decided that college was not really for him and instead followed his passion for sound. He walked into the rental desk at Bridgewater Sound and simply asked, "If I want to get a job in this business, how do I do it?" He was put on the part-time call list and immediately started working some gigs. However, as it was part-time and he needed a steadier job, he worked during the day for a computer refurbishment company and took sound assignments from Bridgewater at night.

The work calls from Bridgewater started out as mostly for election campaign press events that involved a microphone mixer and a press box. He usually operated as a "one-man-show," interfacing with the candidate's press team and the broadcasters to ensure that everything sounded good. He found that he could relate well to these events as they were mostly spoken word and similar to his early volunteer work at the church.

> And at some point, Bridgewater had more business and they also saw that I was a reasonably good worker. So I ended up becoming full-time for them, relatively speaking. I'd work in the shop and then go out and do shows as they came in. We would do things besides the press events for those campaigns that inevitably lead to connecting up with the White House Communications Agency. Plus at some point a candidate wins, then usually runs for another office and brings all the same people along with them. So it was really interesting to see who and what all went along with the candidates.

With the growth of Bridgewater Sound and various campaign-related events, it became obvious that they needed to provide more than just audio. Bridgewater shifted to become a full production company providing lighting, sound, staging, and more. Kelly was able to expand his role to that of a project manager and again found that knowledge and experience from his past were valuable for his current role.

> I learned what was important to a lighting team, to the staging folks, to the logistics team, and how to interact with all of them. In addition, I had the opportunity to cooperate not just with the campaign, but also the Secret Service. During this time I really learned how everybody's interests have to be represented and addressed for each production.

As Bridgewater grew to become one of the primary production and audio companies in the Chicago area, it also provided additional opportunities for Kelly. He would work as an A2 on many other events, such as musical festivals at Grant Park and various music performances at the Petrillo Music Shell. He credits the more experienced mixers and system engineers on the crew for "putting in a good word" with the owner, which opened up an opportunity to mix at the auxiliary stage of the Chicago Jazz Festival in the mid-90s.

> This then led to another growth in my career in a roundabout way. I was invited to help with fireworks on July 3 at Taste of Chicago. At the time the synchronization was via a two-way radio keyed up with the music and the firing count. Because of the broader services Jay Bridgewater would provide, I was able to learn more about two-way radios and RF technology. I instantly knew that wireless technology was going to be very important and that I was really drawn to it.

Seeing the opportunity, Kelly dove right in and began studying all that he could about RF technology. He read books, attended HAM radio conventions, took seminars, and connected with staff at Shure and Sennheiser. Around the same time Bridgewater began working with local television broadcasters for shows that taped in Chicago, such as *The Jenny Jones Show*. They were renting wireless microphones to the production company and Kelly was able to apply his recently learned knowledge and desire to experiment. He was assigned the *Jenny Jones* account and began working with their audio team (which included Mark Harper, Mike Cunningham, and Mike Aiello), supporting their use of wireless mics.

> When *The David Letterman Show* worked with Sennheiser to provide wireless coverage inside and outside their studio, it seemed to accelerate implementation by other broadcast producers. Today we take ease of antenna coverage for granted. Companies like Professional Wireless Systems (PWS) have made multi-antenna couplers commonplace for many productions.

Mike Cunningham, the show's A1 at the time, would challenge Kelly to find affordable solutions for expanding antenna coverage. The producers and director would request microphone coverage areas outside of the studio; green rooms, hallways—anywhere they could get a camera, it seemed. Kelly had to determine how to operate the wireless systems in multiple spaces and under different circumstances. These exercises allowed him to quickly seek out and build relationships with other individuals and companies in the wireless technology space and learn new concepts.

During this time Kelly also worked as an A2 for sports and broadcast events in the Chicago area. Besides RF, Kelly developed an interest in intercom systems. While viewed by many as a "necessary evil," Kelly saw this market segment as an opportunity.

Embracing tasks that others don't enjoy gave me an opportunity to work on more projects. To this day, I still enjoy the communications aspect of our industry. Most recently, the pandemic confirmed both the need and opportunity that intercom brings to the marketplace to enable safe distancing between workers.

In 1994, Chicago was chosen as one of the host cities for FIFA World Cup Soccer finals and the audio and communication design manager was Larry Estrin of Best Audio. Bridgewater Sound was awarded the perimeter speaker systems as well as the press briefing room sound support for both Chicago and Detroit. Kelly was assigned to work with Larry and his team implementing these systems in both cities. This project was another pivotal moment that exposed Kelly to atypical temporary sound systems in unique locations.

His career continued to grow at Bridgewater, but Kelly was looking to expand from regional work to more national and international opportunities. Eventually Kelly left Bridgewater to work for Mario Educate of On Stage Audio (OSA). The company was focused on audio for corporate and large-scale special events. He relocated to Detroit and began working full-time as a project manager, A2, comms tech, or mixer for various shows. OSA also took him around the world with different productions and allowed him to learn how to be adaptable in diverse cultures and with various inventories.

While at OSA, he also had the opportunity to work with state-of-the-art equipment. Eventually he discovered that he did not always need to be the person behind the console. Luckily, OSA offered him opportunities to grow in areas such as intercom and RF technology. While with OSA, he was also reunited with Larry Estrin to help with the Presidential Debates. By this time, Larry was the technical director for the Commission on Presidential Debates. Kelly worked as the project manager on behalf of OSA and was introduced to large-scale intercom systems.

> While we take fiber optics for granted today, in those earlier years we were learning as we went, sometimes painfully. Looking back I wouldn't trade those challenging circumstances for anything. The relationships that grew out of these experiences are worth more to me than any amount of money.

Kelly left OSA in 2006 to become a freelancer and formed his own company named Easy Live Audio. He works now mostly as an audio designer and mixer, although he often brings in others to mix, work as an A2, and implement the intercom systems on shows that he manages.

> Production companies hire me directly and sometimes I do work for vendors as well. But typically my relationship is with the production company. On a given event I may design the system and mix the audio. Sometimes I will hire others to mix for me, but generally I take on the lead design and mixing role. Building the best team for each project is a priority for me.

Figure 40.2 Kelly Epperson's Career Path

Looking back on his thirty-plus-year career, Kelly finds it interesting how his original aspirations changed along the way.

> Early on my two biggest areas of interest were recording and broadcast. I think part of that was because those are the ones that were (and probably still are) the most well-known. My career aspirations at the time were more about which market segments do I want to focus on. But I never really found that the recording niche was a fit for me. Instead live sound was absolutely the place I wanted to be.

His journey provided many lessons that helped to contribute to his success. He recalls reading an article in Mix Magazine while in his 20s about a concert sound mixer, in which the interviewee said: "Sound is what I do and who I am."

> This quote really resonated with me and I sort of adopted that philosophy. However, I also realized that if you're going to be the best at something, or in demand, you're going to have to also be perceived as a great team player. I tried to turn that into a benefit by saying, "I'm going to see the big picture, what's best for the show, but my job is to represent audio on any given project." So at the end of the day, when I'm sitting around the table, if somebody asks me my opinion about things outside of audio, I'm happy to weigh in respectfully. But my job is always to sit at the table and represent audio or the sound of that event.

To this day Kelly is still working on projects and connecting with people from his time with Bridgewater and OSA. He can often be found on projects such as the Presidential Debates, global product announcements, and various broadcast events. And while he is confident that he is on the right path, he never stops learning or trying new things. He is grateful for the insights his experience has given him and the confidence to try new ideas when clients bring a fresh challenge. In fact, he is continuously asking himself, "What are the things you're missing from your knowledge base?" Finding out about what he does not know is often more important to him than what he already knows. He applies this to everything from making decisions to understanding new technology.

He admits that he is at a point in his career where he places a high value on the relationships he has formed with coworkers, manufacturers, vendors, and other production personnel. He continues to build new relationships and also enjoys sharing his knowledge with others. He stresses the importance of always being willing to explain to somebody what you want to accomplish, building consensus when possible, and not being afraid to ask for help. And just as he has done throughout his career, Kelly continues to strive to bring state-of-the-art solutions when needed, while always working to deliver the best overall production.

41 Cricket S. Myers

Freelance Sound Designer

Figure 41.1 Cricket S. Myers

Cricket S. Myers found her love of theater in a roundabout way through physics. Growing up in Michigan, she always found challenges very interesting and something to pursue. So when she discovered that a high school physics class was very demanding, she decided to go to college with the goal of becoming an astrophysicist. Although she was involved in theater during high school, she felt at the time that physics was her calling.

After her first year at Colorado College studying physics and working as a crewmember in the theater, she realized that she *loved* theater and merely *liked* physics. So as a sophomore she changed her major to theater and started down a new path.

> At that point, I was mostly interested in stage management. I had done a little bit of master electrician work and some sound work too. The sound was different than what it is today. It was comprised of reel-to-reel systems, cassette decks, and cutting tape. The good thing is that during this time I got my hands on a little bit of everything. Plus I got out into the real world through a couple of different internships, most of which were focused on stage management.

DOI: 10.4324/9781003188957-48

Eventually Cricket found that stage management was not really for her either. Although she liked the organization and management of things, she was bored working on the same show for 10 to 12 weeks at a time. Luckily for her, another internship opportunity was about to change the course of her career.

While working at the Midland Community Theatre in Texas, she met a man named Eddie Taylor who was in charge of the lighting and sound. She learned about lighting design, general creativity, and even sound effects design through Eddie. This helped her to realize that she wanted to be a designer. After much consideration, Cricket decided to go to grad school for lighting design. She applied to many schools, but an interview with California Institute for the Arts set her straight.

> Coincidently I met with Jon Gottlieb who is the head of sound at Cal Arts. We had what was supposed to be a 15-minute interview that turned into a 45-minute interview. And at the end of it he said, "Listen, I'll be honest, Cal Arts has already rejected you for lighting design; but here's the thing. You're a sound designer and you just don't know it yet. You're not a visual designer. That's why you've never been comfortable in scenic, costumes, or lighting. You're a designer, but you're not visual. So you're a sound designer."
>
> He brought in the dean and I chatted with her for a while. She then looked at Jon and said, "She belongs at Cal Arts. Do whatever you can to get her here." Jon then handed me a form and told me that if I change my application from a lighting application to a sound application, he will accept me on the spot!

She then went back to Texas to complete her internship in Midland with Eddie. Upon telling him of her new sound design focus, he immediately began guiding her on the principles of sound design. She then returned to California to start her education at Cal Arts, where Jon Gottlieb taught her everything he knew over the next three years.

> Sound immediately made sense to me. The technology made sense, because I came from a physics background. So I understood the acoustics and I understood the technology. I understood the way physics affects sound and the way to use that to my advantage. But I also had the creative side of it in that I liked the storytelling. I like the way I could use physics to help tell a story.

Cricket credits her success to a large amount of networking and putting herself into situations to grow and learn. During her schooling and after graduation, she spent a great deal of time meeting and getting to know people in the industry. She would shake hands and give out business cards at every opportunity. She also emailed sound designers and asked to meet for coffee, and called theater companies asking to sit in during tech.

Just out of school she reached out to the South Coast Repertory Theatre and asked if she could come in and observe during preproduction of a show. While there, a sound designer named Michael Roth came in to speak to the audio system programmer and she had the opportunity to meet and talk with him too. That moment allowed her to introduce herself and share her talents. Michael was impressed and asked her to assist him on his next show.

After a successful run at South Coast Repertory, that show transferred to Broadway in 2011. In addition to being Michael's assistant, Cricket was a key part of the production as she was able to translate the audio programming from one system to another using the experience she had gathered at school and through other learning opportunities. She came back to Los Angeles with a Broadway credit that helped her further proceed along her career path.

Due to her aggressive networking and the Broadway credit, Cricket was also assisting several designers including Drew Dalzell and Jon Gottlieb. Not only did they provide her with work, they also introduced her to many other designers. Many of her early design opportunities came from shows that the two of them were unable to do, but suggested her in their place. As her own work grew, so did her reputation through the word of mouth and she began designing more and more shows.

Eventually, Cricket became the Resident Assistant Sound Designer at the Mark Taper Forum.

> This was an amazing opportunity for me. I had the pleasure of assisting many wonderful sound designers as they came through the Taper. I learned so much from every one of them. On top of that, I got paid a great fee for three weeks, and then had six weeks off to pursue my own design work. Knowing that I had a regular paycheck (and health care) meant I could take small shows at theaters around town to really practice my design skills and develop my own aesthetic.

When the Mark Taper Forum shut down for renovations, Cricket seized the opportunity to focus more on her own design work. She started designing musicals as well as plays, and began doing more and more work at the League of Resident Theatres level. Once the Taper finally reopened, she decided not to apply for the head of sound because she wanted to design more than be a house sound engineer. Although it was a great opportunity, she knew that her love of sound was best applied with variety and changes. She did not want to take a job working on long-running shows, so instead she chose to continue with the freedom of working different shows as a freelance designer.

In 2011, Cricket became the first woman to be nominated for a Tony Award for Best Sound Design for a play titled *Bengal Tiger at the Baghdad Zoo*, directed by Moises Kauffman. The story of how she came to work on this show is rather interesting and is a perfect example of how she has built her career by personally taking action.

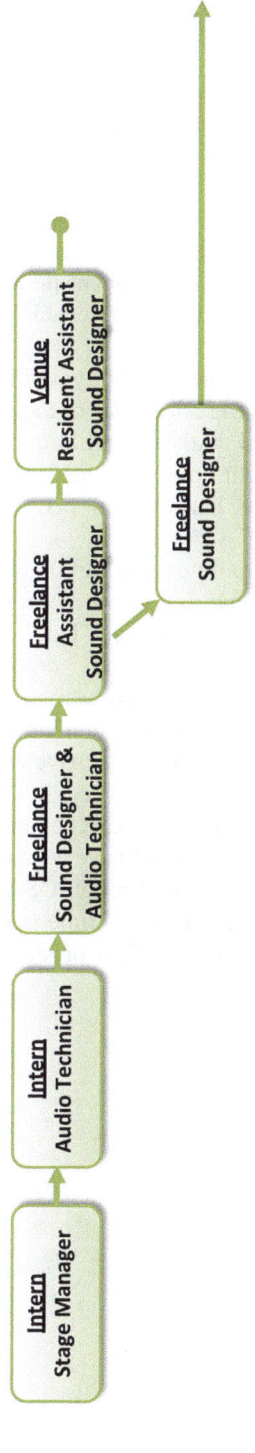

Figure 41.2 Cricket S. Myers' Career Path

It all started when the Taper announced their upcoming season. I saw the play *Bengal Tiger at the Baghdad Zoo* on their list. I thought, "Oh, my God, that looks like an incredible play." So I emailed the production manager and said, "I don't know if you've hired a sound designer for this show yet, but I'm really interested." She wrote back and said they had not found anyone and that she would arrange a meeting for me with the director. I had never been that bold. But obviously, I saw something and went after it.

Cricket's gusto has helped her throughout her career. She credits her continued achievements to a saying that was once spoken to her, stating that true success is based upon three elements: talent, tenacity, and timing.

So you can be really, really talented, but you also have to work hard and you have to go after it. But the third piece of the puzzle is something that's hard to come up with. And that's timing. Sometimes you just have to be in the right place at the right time. And that's always the thing, that there are extraordinarily talented, hardworking people out there that never get that moment.

It is important that you put yourself out there and open yourself up to opportunities. Don't turn anything down, because you never know when you're going to be sitting in the right seat. Don't be afraid to ask questions and be ready to be super flexible.

Remember too that it doesn't matter how many times you ask the orchestra what instruments they are bringing, someone will show up with something they "forgot" to mention, so always have extra mics standing by!

The future holds many exciting things for Cricket and she has not even begun to think of retirement. However, she does have a financial plan in place and is grateful for the pension plan available through USA 829. She looks forward to one day traveling the world and spending time in exotic spaces. Until then, she can be found designing sound for various productions and taking action to further shape her career.

42 Shared Advice

While researching for this book, I conducted a survey of nearly a thousand production personnel. A part of the survey asked the respondents to provide advice they wish to share with others. The results were fascinating, with many focusing on similar themes. What follows are some of the best advice quotes shared from wonderful industry colleagues, broken down into four main categories.

Attitude

- Talk less; smile more.
- Believe in yourself and your abilities, but always listen and work hard.
- Own everything you do, from mistakes to successes.
- Be nice to people, or, at the very least, don't be a jerk.
- Be the sort of person you'd like to work with.
- Keep your head down, keep your work ethic up, keep your mouth shut, and do what you do best.
- Two ears, one mouth.

Design

- If you want to design, design.
- Design should never be about oneself or how you can boost your ego. It is about helping to create a memorable audible experience.
- There is more out there than a theater to design.
- The sound is the easy part; it's the politics that's difficult.

Career

- You have to learn to use a shovel before you can drive the big machine.
- Always be a go-getter. Educate yourself and seek out opportunities to take classes in order to learn and meet people. Don't sit back and wait for things to fall into your lap or to be paid to learn something.
- Don't make career decisions based solely on financial gain. What you choose to do matters more than what you get paid to do it.

DOI: 10.4324/9781003188957-49

- There will always be someone ready to take your place, so don't be complacent.
- If you are a female in the industry, it will be one of the hardest things you will have to do. You have to work twice as hard to be given the same respect as a male walking into the room.
- Some jobs are for rent, some are for art, some are stepping-stones for growth.
- Seek out mentors.

Life

- What makes you happy is the most important. You are the one who has to look at yourself in the mirror every day, so don't make your life-changing decisions based off of what your mentor, colleagues, or peers in the field want you to do. Do what makes you the happiest, because that is what will allow you to enjoy life.
- If you are an enjoyable person to work with, you will get asked to the bar after work. If you get asked, say yes, and don't get drunk.
- You can't judge your own success by measuring it to someone else's. Your path will be unique, but it will not be straight or easy.
- Be exceptional, gain the knowledge, take the training, and get the experience to understand the real and potential risks of any workplace.
- Do what *you* want; it's your life.

A Simple Plan

The survey respondents were also very open to explaining their guidelines for a successful career. As I pored over the answers, it was clear that there were three points shared by a vast majority of the survey respondents. I also agree that these are the key tenements to a prosperous career in the live entertainment audio industry:

1. Show up on time and never be late, no matter your position or stature.
2. Work hard, never stop learning, and always have fun.
3. Be nice and treat others with respect.

While the above can hold true to most industries, they are especially important for success in the live entertainment audio field. Be sure to adhere to these in all that you do, and may you go forth and have a wonderful career.

Index